SCIENCE AND HUMAN EXPERIENCE

~

Does science have limits? Where does order come from?
Can we understand consciousness?

Written by Nobel Laureate Leon N Cooper, this book places pressing scientific questions in the broader context of how they relate to
human experience.

Widely considered to be a highly original thinker, Cooper has
written and given talks on a large variety of subjects, ranging from
the relationship between art and science, possible limits of science,
to the relevance of the Turing test. These essays and talks have been
brought together for the first time in this fascinating book, giving
readers an opportunity to experience Cooper's unique perspective
on a range of subjects.

Tackling a diverse spectrum of topics, from the conflict of faith and
science to whether understanding neural networks could lead to
machines that think like humans, this book will captivate anyone
interested in the interaction of science with society.

~

Leon N Cooper is the Thomas J. Watson Senior Professor of
Science at Brown University and Director of the Institute for Brain
and Neural Systems. He has received numerous awards and prizes
for his scientific achievements, most notably the 1972 Nobel Prize
in Physics (with J. Bardeen and J. R. Schrieffer) for his studies on the
theory of superconductivity.

T0321146

Rembrandt Harmenszoon van Rijn, *Doctor Faust* (1652).

SCIENCE AND HUMAN EXPERIENCE
VALUES, CULTURE, AND THE MIND

Leon N Cooper
Brown University

CAMBRIDGE
UNIVERSITY PRESS

CAMBRIDGE
UNIVERSITY PRESS

Shaftesbury Road, Cambridge CB2 8EA, United Kingdom

One Liberty Plaza, 20th Floor, New York, NY 10006, USA

477 Williamstown Road, Port Melbourne, VIC 3207, Australia

314–321, 3rd Floor, Plot 3, Splendor Forum, Jasola District Centre, New Delhi – 110025, India

103 Penang Road, #05–06/07, Visioncrest Commercial, Singapore 238467

Cambridge University Press is part of Cambridge University Press & Assessment, a department of the University of Cambridge.

We share the University's mission to contribute to society through the pursuit of education, learning and research at the highest international levels of excellence.

www.cambridge.org
Information on this title: www.cambridge.org/9781107043176

First published 2014

A catalogue record for this publication is available from the British Library

Library of Congress Cataloging-in-Publication data
Cooper, Leon N, 1930– author.
[Essays. Selections]
Science and human experience : values, culture and the mind / Leon N Cooper.
pages   cm
Includes bibliographical references and index.
ISBN 978-1-107-04317-6 (hardback)
1. Science – Social aspects.   I. Title.
Q175.5.C664   2014
303.48′3 – dc23   2014012953

ISBN   978-1-107-04317-6   Hardback

To

Michael, Liam, Nico and Manisha
May your world be a happy place

CONTENTS

～

Preface xi

Acknowledgement xiii

Part One · Science and Society

1 · Science and Human Experience 3

2 · Does Science Undermine our Values? 24

3 · Can Science Serve Mankind? 37

4 · Modern Science and Contemporary Discomfort:
Metaphor and Reality 41

5 · Faith and Science 57

6 · Art and Science 59

7 · Fraud in Science 64

8 · Why Study Science? The Keys to the Cathedral 68

9 · Is Evolution a Theory? A Modest Proposal 70

10 · The Silence of the Second 73

11 · Introduction to *Copenhagen* 76

12 · The Unpaid Debt 79

Part Two · Thought and Consciousness

13 · Source and Limits of Human Intellect 87

14 · Neural Networks 109

15 · Thought and Mental Experience: The Turing Test 114

16 · Mind as Machine: Will We Rubbish Human Experience? 124

17 · Memories and Memory: A Physicist's Approach
to the Brain 132

18 · On the Problem of Consciousness 148

Part Three · On the Nature and Limits of Science
19 · What Is a Good Theory? 155
20 · Shall We Deconstruct Science? 158
21 · Visible and Invisible in Physical Theory 170
22 · Experience and Order 177
23 · The Language of Physics: On the Role of Mathematics
in Science 188
24 · The Structure of Space 198
25 · Superconductivity and Other Insoluble Problems 218
26 · From Gravity and Light to Consciousness: Does Science
Have Limits? 225

PREFACE

~

The essays in this collection are taken from articles and lectures that I have written or delivered over many years – some of which I hadn't looked at in quite a while. Putting them together I am struck by the recurrence of certain themes. This shouldn't be surprising since I have been thinking about these subjects for a long time. They occur in somewhat different contexts and reflect the evolution of my thoughts on the relation of science to other human activities.

Since some of the articles were written years ago, I have occasionally included footnotes to update matters where the situation has dramatically changed; but otherwise I have pretty much left things the way they were because that was the way I thought about them at the time.

What is more problematic is that in the originals there are paragraphs and even sections that are repeated from one article to the next. As with itinerant actors and musicians over the centuries, each performance is pieced together from those that have come before – self-plagiarism. But each is presented in a new package and sometimes the package is as interesting as the content.

So I've deleted extended repetitions; also in some of the essays I have excised sections that, in my opinion, don't contribute to the main line of thought and, sometimes, are discussed in other essays.

However, in some cases a few paragraphs are so integral to the argument that I have decided to leave them even though they do repeat what was said before. I imagine that many readers will pick among the essays so that it seems reasonable to include some repetitions to make each complete by itself. For those intrepid few who begin with the first and read through to the last, I can only express my apologies and counsel a willingness to skip the paragraph or two that they have read before.

What we have here, therefore, is a collection of my thoughts as they were and as they have evolved over many years. I've tried to put them into a sequence that makes some sense but, inevitably, one does jump from one topic to another. I hope that these essays illuminate some of my thinking on a variety of subjects concerning the meaning of science and its relation to other human endeavors.

ACKNOWLEDGEMENT

~

I would like to express my appreciation to Nicholas Gibbons, Lindsay Stewart, and the other members of the editorial team of Cambridge University Press for their assistance in editing and preparing this manuscript. I would also like to thank Pete Bilderback for his invaluable editorial, administrative, and organizational efforts as well as for his help in designing the cover.

PART ONE

Science and Society

1

Science and Human Experience: (Mephistopheles Is Alive and Well and Living in the Space Age)

One does not often find written today "The great object of science is to ameliorate the condition of man, by adding to the advantages which he naturally possesses." Has science failed us? If so, how?

This essay is based on a talk given for the Third Tykociner Memorial Lecture at the University of Illinois, Urbana, in 1976. The lecture series, named in honor of Joseph Tykociner, focuses on the relationship between science and art.

1

Professor Tykociner was concerned with what is common to the arts and the sciences as well as the interplay between these very different areas, a subject that has always intrigued me. Whether what is common is more important than what is obviously different will probably be argued about for ever. However, the sharp distinction between the sciences and the arts that has been recently much talked about may not be that well posed. We might as well ask, considering the vastly different techniques employed, whether what the musician, painter, and playwright have in common is more important than where they differ.

I believe that there does exist a deep commonality of purpose among the arts and the sciences. But you will have to decide for

3

yourselves whether this is significant enough – considering their many differences – to be worth emphasizing.

I have always viewed science as rather closely related to other things we do, an elaboration – perhaps an outgrowth – of what all of us do normally in the course of our lives, rather than being separated from other activities by a special method or rationale known to and practiced only by certain chosen high priests. If I were to convey the attack and charm of science in a few words, I would remind you of the great detective: Sherlock Holmes.

Recall his attention to the facts: "What do you think, Holmes?" asks Watson. "The data are not yet sufficient," Holmes answers. For he must separate what actually is from all that might be. But then the most magical stroke. What seem disparate and disconnected events, through the power of his intellect become an ordered and rational whole. How delighted we are. It is the fulfillment of a primeval urge – that the world succumb to the mind.

Suddenly painter, storyteller, magician, detective, and scientist are one. Man has imposed his will on the chaos and disorder of experience.

But this ancient dream has become for some a nightmare. The early summer joy of discovery has given way to a winter of discontent. In spite of the evident and vast benefits showered upon us, we hear increasing complaints about the evils of science and technology. The two words have become so coupled it seems possible that some future dictionary, following modern tradition, might accept them as interchangeable.

Each age has its myths; for ours, the myth of workers alienated from the owners of productive machinery has been replaced by that of individuals alienated from reason and society, and from the structures that have made possible their successful entry into the world and their later survival. Yet, although such views have received immense publicity, I don't believe they are really shared by most people, who are simply too sensible to be swept along with seasonal intellectual fashion. Still one would not likely find it written today that,

The great object of science is to ameliorate the condition of man, by adding to the advantages which he naturally possesses.[1]

And there is no doubt that recently we have seen a particularly violent turn, not only against science but also more generally against what we might call rational argument and analysis. The problem is surely no longer one of social caste:

Well, if you're like me, you loathe all science and mathematics...

says Scott Fitzgerald's Monsignor Darcy and young Amory Blaine, Princeton bound, nods in vehement agreement:

Hate 'em all. Like English and history.[2]

Neither is it likely to be the hard work required to learn the complex language of science, nor the patience required to undergo the long period of initiation before one can begin to do things that are interesting and related to our ordinary concerns. In spite of a certain current preference for instant gratification, students still study foreign languages, music, and a variety of subjects equal in difficulty to the sciences.

This is a period of heightened awareness of the effects on nature of increasing population and industrialization. We have lived through a generation of unnecessary war, a decade in which sophistry and plain lies in high places have undermined our faith in what is said. Science, become institutionalized and bureaucratic, shares the faults of other institutions and has become equally vulnerable to criticism. The public and private behavior of scientists reveals them as ordinary human beings, which apparently

1 Webster, J. (1808). *Elements of Natural Philosophy*, Philadelphia: B. and T. Kite, Fry and Kammerer, p. v.
2 Fitzgerald, F. S. (1920). *This Side of Paradise*, New York: Penguin Books, p. 30.

surprises – even shocks – especially those who feel left out of the priesthood and view it with both fear and envy. Perhaps it should not startle us that this has led to suspicion that often goes as far as rejection of rational argument, especially among those who are young, pure, and most affected by the errors of their elders.

Yet accepting all of this, there is a certain irony (and I am sure I am not alone feeling this way) that people will complain about the evils of electricity in the uniform warmth of their homes while keeping all the lights on, the evils of medicine fully expecting that their children will survive early childhood and while taking penicillin to cure a strep throat. They will complain about the dangers of research in physics while listening to their transistor radios and enjoying in the privacy of their homes at a flick of the finger whatever music they desire. And I am sure that in the future when we no longer die of various dread diseases that terrify us today, and automobiles glide safely, odorlessly, and silently on cushions of air, we will continue to hear similar complaints.

That present middle-class comforts were not always available is obvious; that they were desired is perhaps not so well known. We won't mention the lack of plumbing in Versailles, nor the famous winter evening when the wine on the royal table froze. In 1887 Edward Bellamy wrote ("looking backward" from the Utopia he was visiting):

... if we could have devised an arrangement for providing everybody with music in their homes, perfect in quality, unlimited in quantity, suited to every mood, and beginning and ceasing at will, we should have considered the limit of human felicity already attained, and ceased to strive for further improvements.[3]

Now we have indoor plumbing and central heating; we even have music in our homes but we are still not happy. Certainly not happier on the average than in times past. And if we are not happy

---

3 Bellamy, Edward (1888). *Looking Backward: 2000–1887*, New York: Random House, p. 90.

who or what is to blame? Didn't the World's Fair exhibit promise us a new world of the future, a world full of gadgets that did all the work at the flick of a finger? And weren't the human figures who populated that world always smiling?

You will not be misled, I am sure, into believing that I am trying to convince you that we live in a world where all is good and where "science and technology" lead us ever onward and upward. However, because the simplistic notion of continual progress is not an adequate description of history seems to me insufficient reason to replace it by an equally simplistic notion of continual decline. History has its ups and downs. All of us can look with nostalgia to some feature of the early or distant past that we wish could be retained. It is true of course that things look better in the past – if we can become sentimental about the Depression era, or consider the fifties exciting there must be something about memory that improves the quality of events.

Recently, Victor Weisskopf, a humane and learned man and certainly among the eminent scientists of our time, in a talk titled "The frontiers and limits of science" stated optimistically that "at least potentially, science can justifiably claim the ability to understand every observable phenomenon." This statement was of course qualified: "...there are many phenomena...that we do not yet understand. But...it is reasonable to predict that man will eventually understand all of nature scientifically."[4]

I was struck, however, with Professor Weisskopf's concern about the limits of science:

A Beethoven sonata is a natural phenomenon which can be analyzed physically by studying the vibrations in the air, as well as chemically, physiologically and psychologically by studying the processes at work in the brain of the listener. However, even if these processes are completely understood in scientific terms, this kind of analysis does not touch what

4 Weisskopf, V. F. (1975). The Frontiers and Limits of Science, *Bulletin of the American Academy of Arts and Sciences*, **28**(6), 15–26.

we consider relevant and essential in a Beethoven sonata – the immediate and direct expression of the music. In the same way, one can understand a sunset or the stars in the night sky in a scientific way, but there is something about experiencing these phenomena that lies outside science.

Now I know no one who would claim that a Beethoven sonata or for that matter a Judy Collins song could somehow be replaced by science. Yet while I am not in disagreement when Professor Weisskopf says: "there is something about experiencing these phenomena that lies outside of science," I find striking the implication that what has failed us is science.

If there is a sense in Victor Weisskopf that science is "limited" and somehow has failed us, how much more violent and explicit this is in Theodore Roszak. Writing in *Daedalus*,[5] Roszak, whom one would not call an admirer or a lover of science, invokes Mary Shelley, the child author of *Frankenstein, or The Modern Prometheus*: "A girl of only nineteen...she joined the ranks of history's great mythmakers. What else but a myth could tell the truth so shrewdly, capture definitively the full moral tension of this strange intellectual passion we call science?"

"Asked to nominate a worthy successor to Victor Frankenstein's macabre brainchild what should we choose from our contemporary inventory of terrors?" Roszak ask rhetorically. No lack of candidates: "The bomb?...the behavioral brain washer? The despot computer? Modern science provides us with a surfeit of monsters, does it not?"

Not all scientists, Roszak admits, are mad doctors. He realizes that there are those who champion "a science for the people" and he writes "in full recognition of how the wrong-headed power elites of the world corrupt the promise of science." But all of this is preliminary because, he continues:

5 Roszak, T. (1974). The Monster and the Titan: Science, Knowledge, and Gnosis, *Daedalus*, Summer 1974, 17–32.

I have another monster in mind that troubles me as much as all the others – one who is nobody's child but the scientist's own and whose taming is no political task. I mean an invisible demon who works by subtle poison, not upon the flesh and bone, but upon the spirit. I refer to the monster of meaninglessness. The psychic malaise. The existential void where modern man searches in vain for his soul. 'Further' ... it is science which has made our universe an unbounded theater of the absurd. ...

And so on.

To most this will seem a trifle exaggerated. After all Frankenstein monsters have been created by bureaucrats as well as scientists, and the existential void may reflect personal as much as cosmic deficiencies. However, if so eminent a scientist as Weisskopf and so harsh a critic as Roszak intersect anywhere in their doubt, perhaps it is worth considering what might be troubling them.

I would like to discuss, then, why it is that science, which on its own terms has really been successful beyond the dreams of its founders, has come to be feared, distrusted, and so savagely attacked. What is the source of that unease expressed by scientists themselves? Has science really failed us? If so, how?

2

We are not the first to have experienced this disenchantment. Let me remind you of Dr. Faust, eminent scholar and renowned professor. In Goethe's masterpiece we meet Faust first in his chamber, learned and old:

> ... Look at me.
> Years wasted grinding through philosophy,
> Slaving over medicine and law,
> Learning everything – dear God even theology.
> Now here I am with all this lore
> Poor fool no wiser than before.[6]

6 All quotations from Goethe's *Faust* are taken from or based on the prose translation of Barker Fairley, University of Toronto Press, 1970. An occasional jingle may be mine.

He is one of us: tenured professor, rich in graduate students whom he leads "by the nose this way and that, upstairs and downstairs." He knows everything, but what does it amount to? Something is missing; he turns to magic, conjures up the devil: not Lucifer himself but Mephistopheles – sort of a deputy. Some portions of his story such as his compact with the devil and his seduction of Marguerite are well known. Less so his later adventures. In the end (and the end is a very long way off) in spite of the compact he is saved.

As is often the case, legend has its roots in fact.[7] The actual person we now know as Faust was born around 1480 in the small village of Knittlingen in the southwestern corner of Germany. His given name was George; he was rechristened John, and Goethe, for reasons we don't know, calls him Henry (Heinrich). His last name was Faustus, possibly nothing more than the fairly common name "Faust" with a Latin ending. His title Doctor.

During his life he wandered through many towns across German-speaking territory pursuing his career as a magician. He had already attracted considerable attention as a practitioner of magic when he was about 26 years old. An authority as reputable as Luther's associate, Philipp Melanchthon, tells us that Faust studied magic in Poland at Cracow University. (Magic, we discover, was a university subject at that time – the old, new curriculum.)

He was a braggart: boasting for example that he knew more alchemy that all other alchemists put together; claiming, in the years 1525 to 1527 when imperial troops were fighting in Italy, that he had produced their victories by remote control. And a performer. When he lectured on Homer at the University of Erfurt in 1513 he is said to have enlivened his classes by producing the

---

7 These "facts" are based on an account given in Charles E. Passage's introduction to his translation of *Faust*, Indianapolis: Bobbs-Merrill Co., 1965. I would like to express my appreciation to my colleague, Professor Karl Weimar, for directing me to these two translations of Faust.

Homeric heroes alive before his students' eyes. And he must have convinced someone. The minutes of the city council of Ingolstadt for June 17, 1528 indicate that Faust was ejected from that town but only after the council had forced him to promise not to take revenge on them.

He grew famous, lived riotously; crowds flocked to him. When a Franciscan monk attempted to change his ways, Faust informed the monk that he had made a compact with the devil. Assured that God's mercy was infinite and that the Franciscans could perform a mass of purification, Faust, we are told, replied "Mass this, Mass that, the devil has fairly kept what he promised me and therefore I intend to keep fairly what I have promised and signed away to him."

Then, to confirm everyone's suspicions, when he died (in Staufen, another town of the same region, possibly in 1539), his corpse was found face down – a sure sign of his dubious associations. The fate of his black dog – whose spirit, as was well known, alternated between the form of dog and butler – is not recorded.

We have here sort of an early sixteenth-century Uri Geller, a wandering magician and charlatan who obviously captured the attention of the locals. How he managed to escape the stake, we are not sure.

Within a generation Faust's story had been put onto paper in student's Latin by an author who remains anonymous. The German version, called the Faust book, was published by Johann Spies of Frankfurt-on-the-Main, Goethe's native city, in 1587; from this version came all the subsequent Fausts.

The legend was taken up by Christopher Marlowe in England in *The Tragical History of the Life and Death of Dr. Faustus*, which appeared at the very beginning of the seventeenth century. In Marlowe's conception Faust emerges a nobler and more imposing figure than ever he was in real life. Here George, the village magician, assumes Faustian proportions:

> Had I as many souls as there be stars,
> I'd give them all for Mephistopheles.
> By him I'll be great Emperor of the world,
> And make a bridge through the moving air,
> And pass the ocean with a band of men;
> I'll join the hills that bind the Afric shore,
> And make that country continent to Spain,
> And both contributory to my crown.
> The Emperor shall not live but by my leave,
> Nor any potentate in Germany.

The legend traveled back to Germany. Gotthold Lessing in 1759 produced his version of Faust, of which it is not clear that anything more than a fragment ever existed. To the original themes, however, Lessing added the idea of Faust saved. At the end of the work a celestial voice was to cry:

Don't gloat. You have not triumphed over Man and Knowledge. God has not given Man the noblest of impulses only to make him eternally miserable.

By Goethe's pen the wandering magician achieved his ultimate meaning and grandeur. Goethe worked on his Faust intermittently during his entire life. He wrote the Marguerite portion when he was young, and completed the final volume between his 76th and his 82nd year. The last composition was in the midsummer of 1831 when he was 82 years old. The manuscript was sealed at the end of August and left with the instructions that it should be published only after the author's death. Goethe died the following spring and within the same year, 1832, Faust, the second part of the tragedy in five acts, appeared in print.

Since that time, Faust's story has been told and sung countless times. The most famous element of the legend – the compact with the devil, coming from Faust's original statement: "Mass this, Mass that ..." has been repeated in every version from Marlowe's Faustus to such Broadway productions as Damn Yankees.

But what will I give you in return?

Faust demands of Mephistopheles in the Gounod's opera.[8]

Mephistopheles: Almost nothing: Here I am at your service, but below you will be at mine.
Faust: Below?...
Mephistopheles: Below.

In Goethe's poem Faust is more relaxed:

Faust: And what have I to do in return?
Mephistopheles: There's plenty of time for that.
Faust: Not a bit of it...State your terms.
Mephistopheles: I pledge myself to your service here and will always be at your beck and call. If we meet over there, you can do the same for me.
Faust: Over there is no concern of mine...

When Mephistopheles asks: "Could I have a word in writing?" Goethe's Faust mocks him:

So you want it in writing, do you, you pedant?...What do you want, you devil? Bronze, marble, parchment, paper? Shall I write with a style or a chisel or a pen? Take your choice.

Mephistopheles insists: "Any scrap of paper will do. Only you must sign with a drop of blood."

Faust: If this really satisfies you, we'll go through with the tomfoolery.

Gounod's libretto emphasizes pleasure:

Faust: *A moi les plaisirs,*
     Les jeunes maîtresses!
     A moi leurs caresses!
     A moi leurs desires![9]

---

8 *Faust*, music by Charles Gounod; libretto based on Goethe's poem by Jules Barbier and Michel Carre.
9 "Let me have pleasure, / Young mistresses, / Let me have their caresses, / And their desires!"

For Goethe, Faust's aim is somewhat broader: "it's not a question of enjoyment...", he explains to Mephistopheles. "I mean to expose myself to all the pain and suffering in the world. No more. All that is given to humanity to experience I desire to experience myself, the heights and depths of it..."

In all that is given to humanity to experience, there is nothing that beats palpitating young love. And of course it is young love that Gounod, end of the nineteenth century romantic, has chosen to capture. Mephistopheles calls on night to shut out remorse:

> O nuit, étends sur eux ton ombre!
> Amour, ferme leur âme aux remords importuns!
> Et vous, fleurs aux subtils parfums,
> Epanouissez-vous sous cette main maudite!
> Achevez de troubler le coeur de Marguerite![10]

Faust and Marguerite are now alone in her garden. What they feel can only be expressed in song. We live in what may come to be called the era of the super-cool. But hopefully, behind each indifferent façade there is the memory of at least one such night. But it is not easy. Marguerite, frightened, begs Faust to leave. Faust protests but agrees. However, he is stopped at the garden gate by his ever-attendant servant:

Mephistopheles: Learned doctor, you badly need to be sent back to school.
Faust: Leave me.
Mephistopheles: Just listen a minute...She is opening her window.

Marguerite sings to the stars of her love; Faust rushes back. And Mephistopheles, in one of drama's cruelest moments, fills the night with his mocking laugh.

---

10 "Oh night, cover them in your shadow! / Love, close their soul to inopportune remorse. / And you, subtly fragrant flowers! / Blossom under this accursed hand, / To unsettle the heart of Marguerite."

Marguerite has fallen. Now everything goes wrong. She is pregnant. Her brother is killed by Faust. She bears and drowns Faust's child and is condemned to death. Goethe wrote this portion of the poem early in his life, and it possibly reflects an unhappy love affair. Still it was dramatically acceptable. History has its ups and downs. We are told over and over how bad things have become today. But it is not only the plumbing that has improved in the last two hundred years. Hopefully, we are rid of the mentality that demanded she be punished so cruelly.

Her reward will be in heaven. Interrupting a Walpurgis night orgy Faust, dragging Mephistopheles with him, returns to the prison cell to try to save Marguerite from execution. She refuses to go with him. We are familiar with her final radiant theme. She dies. "Judged," cries Mephistopheles. "Saved," sings the angelic chorus. (We know she is saved for we can hear the harp.)

But Marguerite was innocent, pure of heart, seduced with the assistance of Satan himself – surely deserving of God's mercy. What shall become of Faust? For this we must travel a long road. The second part of Goethe's epic is as complex a work as has ever been written; in the course of this, Faust is joined by Mephistopheles in a series of adventures during which he can truly have said to have been exposed, in his own words, "...to all the pain and suffering in the world...All that is given to humanity to experience...".

He is involved in politics and military campaigns. He meets classical and romantic heroes. He has a child, named Euphorion, by Helen of Troy. (The child possibly represents Byron, personification of Romanticism – Goethe's attempt to unite the Greek ideal with the Western spirit.) For the full richness, I must refer you to the poem. However, you might be amused by one foray into high finance. In the fourth scene of the second volume, Faust and Mephistopheles are in the court of the emperor when various ministers enter to announce a miracle. The emperor's debts have been paid. A paycheck has been issued and the army is loyal again. The entire realm is full of joy. Everyone is rushing about spending the new paper money that has just appeared in circulation as fast as possible. The treasurer explains that the emperor himself is responsible, for the night before, dressed as Great Pan, he has signed a master note from which thousands of copies have already been run off. The paper money is redeemable in gold that lies in the emperor's soil – as yet unmined. Who advises Arthur Burns?[11]

In the end, enraptured by a vision of a Utopia, Faust utters the fatal words:

Now I could almost say to the passing moment: Stay, oh stay a while, you are beautiful…I now taste and enjoy the supreme moment.

And now the great struggle between God and Satan over the soul of Faust reaches its climax. Gounod's opera has long since ended. It is Boito who renders for us Faust's glorious triumph. He is raised to the heavens among sounding trumpets and exulting angels while Mephistopheles whistles below.[12] Indeed,

God has not given Man the noblest of impulses only to make him eternally miserable.

3

We are left with the uneasiness that, in my opinion, is the heart of that widely felt emotional rejection of science. Above all, it is the fear that science may deny us our uniqueness and individuality, thus providing the intellectual basis for that unloved, unwanted, bureaucratic, and computerized world of 1984 – that science, having no place for our humanity, comes between ourselves and our own feeling.

It is this I think that is troubling Weisskopf when, speaking of the Beethoven sonata or the sunset, he says uneasily: "there is

---

11 Or Ben Bernanke and Janet Yellen.
12 *Mefistofele*, music and libretto by Arrigo Boito.

something about experiencing these phenomena that lies out-side science." For Weisskopf surely agrees that not only the phenomenon of the sonata, but also the phenomena of the interaction of that sonata with the listener – which encompasses the listener's sense and experience of the sonata can be understood in terms of physical phenomena. What is it then that lies outside of science? "...this kind of analysis," Weisskopf says, "does not touch what we consider relevant and essential...the immediate and direct expression of the music."

We must, of course, distinguish science as observation and explanation from experience itself. Knowledge and understanding, even understanding of experience in terms of "physical phenomena," is not a substitute for the having of the experience – as Faust tells us clearly in the first few lines he speaks. For experience involves nerve endings as well as the brain – whereas understanding engages the brain alone. This, it seems to me, is neither a limitation on science, nor is it threatening.

But what is threatening, even frightening, comes from the suspicion that science, in addition to making it less fashionable to believe, has made it less fashionable to experience or to feel. For where is feeling in the great structure of science? Will the Laplace of 1984 tell us with a sneer that he did not need that hypothesis?

When it comes to feeling, academia generally becomes uneasy. I recall a meeting concerning educational policy attended by representatives of the humanities and sciences. It was a scientist who suggested that in the humanities a student's emotions might be engaged – an intriguing, if unintentional revelation of his state of mind. The response of the humanities representative was equally instructive. He quite violently took issue. The humanities, he insisted, are not a domain where emotion and feeling are predominant. They are as scholarly, analytic, and emotionless as is science. Quite right. Where then is emotion in our systems?

Scientists, arrogant with success have not made matters easier. Listen to the supreme materialist Holbach:

The universe is a vast assemblage of everything that exists, presents only matter and motion. The whole offers to our contemplation nothing but an immense and interrupted succession of causes and effects.[13]

How these words "only" and "nothing" sneer at our pretentions. It is the same sneer that informs us that we are constructed of only $0.90 worth of elements (the price must be higher now). The implication is clear: "What makes you think you're so special? – What right do you have to feel?"

The position is insulting and degrading. But it is also incorrect. For science has no need to assume feeling. Feeling exists. It is a primary fact of our experience, as immediate as any other. Rather science must explain how it is that we feel. If science is to remain within the present framework of its assumptions, it must show us how a creature constructed of ordinary materials that obey "physical laws" can come to have that unique sense of himself which is the possession of all humans and possibly some animals as well. We thus are led to a major remaining mystery – a frontier scientific and human problem: How is it that a machine can feel?

There is precedent that such a question can be answered. Statistical mechanics, which begins with atoms or particles distributed at random but obeying the laws of motion, succeeds in finding a combination of quantities that can be identified with what macroscopically we call temperature. Molecular biology and chemistry have presented us with an entity that can be identified with what previously was called the gene. And in the greatest achievement of nineteenth-century physics, Maxwell identified what we know as light as an electromagnetic wave.

It could turn out, however, that we will not be able to construct an entity such as ourselves from ordinary materials following physical laws. When confronted with this possibility scientists – being attached to their science – sometimes fall into the cardinal error of intellect. They deny what is because it does not fit into their system

13　D'Holbach, P. (1770). *The System of Nature*, trans. Samuel Wilkinson (1820).

of the world. Viewed properly, we would have to regard such an eventuality as a discovery of revolutionary properties. A discovery that rather than undermining our identity would require a fundamental modification of physical laws in order that they yield the identity we in fact have.

I, personally, see no indication that this last eventuality will actually come to pass. There have been many attempts, recent and not so recent, to assign consciousness and intelligence a special place apart from ordinary objects. In Spinoza's words:

They appear to conceive man to be situated in nature as a kingdom within a kingdom: for they believe that he disturbs rather than follows nature's order ... [14]

It seems more likely to me that, just as prior grandiose assumptions concerning vital forces and organic materials, these too, in the near future, will be quietly laid to rest. Probably things are more straightforward and less threatening than we like to believe. And it might turn out, in spite of the popular opposition between the cold logic of science and the warm spontaneity of art, that science may yet provide us with a concrete realization of a human being as unique and individual as any ever heralded in poetry or song.

4

Let us return to Faust. What is the poet trying to say? We cannot, of course, attempt to do justice to the many learned commentaries. But (as Kafka might say) all of the commentators agree on certain points: Faust knew everything that could be known and was not content. Through Mephistopheles he achieves "All that is given to humanity to experience ..." In the course of this he becomes further and further separated from Mephistopheles. And in the end he is saved as though in Lessing's words:

14 Spinoza, B. (ca. 1677). *The Ethics*, Part III.

God has not given Man the noblest of impulses only to make him eternally miserable.

Somehow "the force that always tries to do evil" has once more done good.

In all of the poem's richness and diversity, Faust's gradual but inexorable separation from Mephistopheles is particularly intriguing and suggestive. Faust progresses, in the course of his adventures, from initial exultation, to try everything – to grasp "all that is given to humanity to experience…the heights and depths of it…" to his own esthetic criteria elaborately constructed from classical and romantic models, to his final desire – to create a new Utopian society (to the utter contempt of Mephistopheles) and it is this he thinks he is doing when he is moved to utter the fatal words to the passing moment: "Oh stay a while, you are beautiful…"

The first and greatest triumph of Mephistopheles magic brings Faust youth:

Je veux la jeunesse![15]

sings Gounod's Faust. Youth is wonderful (especially when combined with the wisdom of age); but we must not conclude that Faust must be young to advance his seduction of Marguerite. With Mephistopheles' magic and unlimited wealth Marguerite might as easily have fallen in love with an older Faust. If anything, youthful Faust's inexperience almost ruins the seduction. He must be young again so he can feel what youth feels. Marguerite trembles; Faust trembles with her. They suffer separation. She dies. He feels remorse. They experience all of the disorderly, dangerous, magnificent, passion of young love.

Mephistopheles meanwhile plots. He is the means, the instrument by which Faust is offered that which shapes him. It is the continual requirement of Faust's own sensibility – the development

15 "I want youth."

of that which is most human, through his own experience – which separates him from Mephistopheles and saves him.

His triumph is glorious. It is the victory of God over Satan, good over evil and also the triumph of order over the chaos of raw experience. We exult with Boito as Faust is raised to the heavens to triumphant trumpets and voices of the angelic chorus, while the tempter whistles below.

Passion is wild and full of risk; emotion can be messy as well as dangerous, as Dante discovers when he encounters Paolo and Francesca in the second circle of hell. Perhaps they are better left to the movies. We rather like a reasonable and orderly existence, as we like the comfort of our middle-class homes. Why not? It makes sense. It is safe. We seek that order in our work, our home, and for our children, whom we try so hard to protect. Above all other human activity, science has come to represent order – cold, logical, certain – the order of the inexorable computer, the order of maximally efficient robots which function independent of human desire or need, the order of a world where the inefficiency of passion and feeling have forever been banished.

There is a certain irony in this for, as those of us who do it know, this is mostly science of the textbook. Science when it is being created is robust, disorderly, and often confused. A certain justice too, perhaps, for science has created a stunning ordering of our experience and has helped provide a means whereby we can in fact order our daily affairs. And though we glorify that order, we exult when Faust is saved, at the same time that widespread confusion of the ordering of experience given to us by science with experience itself has led to fear: fear that we may know too much, understand too much, in fact that science may succeed too well. For we are terrified of an order so implacable it would deny the vagary of our desire.

Mephistopheles is alive today, but I confess that I'm not sure he's well. And we need him, poor devil that he is. Our world is so burdened with necessity that feeling and experience seem often

Fig. 1.1 Paolo and Francesca by Gustave Doré.

dispensable luxuries. But his rich bass voice mustn't be lost in the shuffle of papers and the relentless whir of computer tapes. We need him because each one of us must pass by the route of our own experience, dangerous and disorderly as it is. To savor fully

the sounding trumpets and the heavenly chorus, we must somewhere on the path have sung, if in a halting and untrained voice –
even if early designated hummer or listener, a duet or two with the
fellow who whistles.

2

Does Science Undermine our Values?

Is science compatible with our values, or is it opposed to them? Does science have a value system of its own, or is it value neutral? Why is the image of science as creator of Frankenstein monsters so persistent despite the many material benefits science has bestowed upon us?

This essay is based on a talk given in 1977 at Brown University that focused on the role of values in liberal education and was later published in the Brown Alumni Magazine.

1

In a world full of noise about the evils of science and technology, a world overflowing with PCBs, nuclear wastes, carcinogenic agents of every description, a world threatened with ozone depletion, recombinant DNA, to mention a few, I am asked to write on the role of values in what Theodore Roszak, a not too friendly critic, labeled "this strange intellectual passion we call science." Presumably what is expected is something in the vein of the 1808 *Elements of Natural Philosophy*, where we are informed that: "The great object of science is to ameliorate the condition of man, by adding to the advantages which he naturally possesses."

Call it wide-eyed innocence, but I believe most people are not taken in by much of the nonsense they are subjected to and, though confused and troubled, are too sensible to be swept along with seasonal intellectual fashion. Most of us realize that although we can look with nostalgia to some feature of the early or distant past that we wish could be retained, though we no doubt could do without many of the idiotic plastic gadgets that are showered upon us, yet finally we can't feed and house a population of four billion with even a modicum of middle-class comfort by hunting with bow and arrow and relying on organic agriculture.

Whether or not we are willing to trade our comfort for nature virgin I'm not sure, but it seems to me that often the loudest voices against everything are those whose houses are already the warmest. We discount completely that in the not too distant past what we now accept as normal was not available and that people were not necessarily content in blissful ignorance. Let's not mention the lack of plumbing in Versailles, or the famous winter evening when the wine on the royal table froze. After all, would we trade our innocence for a bathtub or a little heat.

You will not believe I am trying to convince you that all is well in this sorry world of ours. We are all aware of the enormous problems that accompany the industrialization that supports our increasing population. And we are aware of the problems of providing life support for concentrations of millions of people in postage-stamp areas of real estate. But, when we consider the harm or value of new technologies, we might remain just a bit calmer and less paranoiac. No one proposes "scientifically distributing" three pounds of plutonium so as to produce, according to Ralph Nader's estimate, "three billion doses of cancer."[1] An equally "scientific distribution" of nature's supply of snake venom would also do us all in (not to mention aluminum, arsenic, lead, petroleum . . . and so on).

All of this appears to me to be more or less obvious. Yet when confronted by difficult practical judgments we have to make concerning new or existing technologies, an irrational element often

---

1 Ralph Nader claimed "Half a kilogram of plutonium, spread evenly around the world, is enough to induce lung cancer in every person on Earth," in a widely publicized 1975 speech at Lafayette College, and that a pound of plutonium could kill 8 billion people.

dominates. This is an era characterized by skepticism about official pronouncements of any kind (a skepticism well earned, it is true, by official acts of the recent past). But beyond what one might call proper and reasonable distrust, in those discussions concerning the application of science to everyday affairs, there lurks an uneasiness that is troubling enough to fuel the paranoia that seems so often to surface.

I suspect that this is so because, with all of our respect for and admiration of science, many of us share, to some small extent, the feeling of science's bitterest critics. Whether this is directed particularly at science or is a somewhat visceral reaction to all rational thinking, however the image of science creating Frankenstein monsters remains. It is persistent and frightening. Are we afraid that science, like Victor Frankenstein, creates monsters, or, what is even worse, that science is itself the monster destroying our values and our human worth?

Science is powerful and, like Mary Shelley's monster, moves with a directive of its own, oblivious to human value or desire. If there is an object to one more essay on this subject, it is to look more closely at this monster so that we might regard him without terror.

2

Newton perceived a flaw in the clockwork grandeur of his system. He was not able to calculate the effects of forces the planets exert upon each other, and so could not be sure that the planetary orbits would be stable. It is said he suggested that these orbits might, on occasion, have to be corrected as though each millennium God would adjust his watch, which ran a little slow. By the beginning of the nineteenth century, Laplace believed that he had succeeded in calculating the effects of such perturbations and had demonstrated that the planetary orbits would be stable. Rather than running fast or slow, the watch ran precisely on time. When Laplace was introduced to Napoleon, the emperor asked the author of *Celestial*

Mechanics: "And where is God in your system?" Laplace, we are told, answered: "*Je n'ai pas eu besoin de cette hypothèse.*" ("I did not need that hypothesis.")[2]

We are struck with admiration: how perceptive the emperor was – to have asked just the right question. Perhaps politicians were better briefed in that time. And we are taken aback by the incredible arrogance of this great scientist, which for many is the arrogance of science itself.

But we may miss the clear implication that, had Laplace needed the hypothesis, he probably would have been willing to make it.

In miniature this epitomizes what science is. One does not make an hypothesis one does not need. But further, and often overlooked, there is no hypothesis one will not make if one does need it. What is required is that the hypothesis be well formed; we must know what it means. We must, as Galileo put it, "make what is said depend on what was said before."[3] His teachers of mathematics taught him this method.

Of course scientists don't make revolutionary hypotheses every day. We work within a system that for the moment is accepted. This gives us a sense of the permanence of our assumptions that, however, often is illusory. Even a cursory view of the history of late-nineteenth- and twentieth-century physics shows clearly that physicists have been willing to assume whatever was necessary no matter how revolutionary or shocking.

Little was dearer to the nineteenth-century physicist than Newtonian mechanics and the wave theory of light. Yet, under the pressure of the new atomic phenomena, the assumptions underlying these theories were abandoned. Causality – a concept on which almost all of Western thinking is based – has been modified.

2 This exchange is frequently cited, but most likely apocryphal. It is cited in W. W. Rouse Ball (1908) *A Short Account of the History of Mathematics*, 4th edn, London: Macmillan.
3 Galilei, G. (ca. 1590). *On Motion, and On Mechanics: Comprising de Motu*, Madison: University of Wisconsin Press, 1960, p. 50.

David Hume had suggested that causality was not in the phenomena themselves, but was a concept introduced to order our experience – the kind of philosophical conjecture that elicits knowing smiles from working scientists. But again, to bring order to phenomena in the atomic domain, the quantum theory – twentieth-century replacement for Newtonian mechanics – has taken us from a completely deterministic theory to one in which equal causes do not produce equal effects. Possibly most incomprehensible is Einstein's redefinition of a time interval – a concept almost impossible to communicate, because the everyday definition of time is so embedded in our language and our early training. With all of these developments so recent, it seems surprising that any orthodoxy at all could develop. Yet it takes less than a generation for quantum mechanics, renormalization theory, or whatever to become new gospel.

These revolutionary changes, perhaps because they came at a time when positivism was high fashion, became confused with the idea that physical theories must not contain quantities that are "unobservable." In my opinion, this is somewhat of a misconception. It seems to me that what Einstein, Heisenberg, and those other giants taught us is not that a physical theory cannot contain entities that are unobservable (the wave function, for example, is not in itself observable), but rather that a physical theory is not required to mention or contain entities if they are unobservable. For if something cannot be experienced there is no reason that it must be built into physical theory. It might be possible to build a broader class of theories without it.

It would be unreasonable, for example, to require that theory contain a wave function. In quantum theory the wave function is a means by which the theory is made to yield results that are comparable to observation. Any other theory that yields results equally in agreement with observation would be equally acceptable.

Forty years from now – not to speak of a century in the future – physics is unlikely to have the same shape or to be founded on the same assumptions we make now. We can reasonably expect that currently fashionable assumptions will be abandoned, while unexpected new ones will replace the old.

It is here that some of our problems arise. For there are many beliefs, concepts, or ideas that are dear to us, that we wish to retain even though science does not need them to construct its theories. But science, like Laplace, is strict – it makes no assumptions unless they are necessary.

I have never been content with the often-repeated emphasis on scientific method. Possibly there is some method in science that could be distinguished from the usual madness in everyday affairs, but this seems overdone. I rather think that this so-called scientific method exists less in working laboratories than in antiquated textbooks. What science does – for all its sophistication of technique and subtlety of thinking – is quite straightforward and is duplicated in a great variety of other human activities. First, science observes. Perhaps this conjures up visions of laboratories and devoted workers in white coats. No doubt both laboratories and laboratories containing devoted research workers in white coats exist. But their purpose – to distinguish the world in which we live from all other possibilities of our fantasy or imagination – is as necessary for the detective or the plumber as the scientist. Once the so-called facts are assembled, it is the problem of theoretical science to provide a fabric, a theory, a set of relationships – in short, an explanation that ties together every phenomenon with every other. Again, a task required of the plumber if he is to locate the origin of the leak or the detective if he is to identify the criminal.

As you all know, the explanations it has been possible for us to construct are truly remarkable. And, at this point, I might lead you through the wonders we have created – from the world of elementary particles and atoms, to electricity and chemistry, to DNA and living matter, and so on to conclude that what has not yet been explained we will be able to explain in the future, so that, as Victor

Weisskopf has written: "...man will eventually understand all of nature scientifically."[4]

You have heard this before. Though I share both the optimism and the belief, I would rather emphasize again that this quintessential activity – the construction of explanation – is not all restricted to science. It seems rather more likely that the animals from which we descended have been concocting explanations, even prior to their entry into the sacred classification – human. In his short masterpiece, *Heart of Darkness*, Joseph Conrad tells of the partially educated savage who fires up the vertical boiler that powers a steamer going down the Congo River. "He was useful," Conrad says, "because he had been instructed; and what he knew was this – that should the water in that transparent thing disappear, the evil spirit inside the boiler would get angry through the greatness of his thirst, and take a terrible vengeance."

It is difficult not to smile. That is certainly not science, we say. Perhaps not, but it is explanation. And the desire and need for explanation is older than what we call science; I would say that the psychological necessity for explanation is one of the early prerequisites for science. For humans refuse to accept a world without order; among the most ancient and honored of human occupations is the creation of order. And, if we look honestly into ourselves, we feel our remote kinship with this wild, passionate, personified world of imagination – and perhaps the faintest trace of a response.

It is a common misconception that science is separated from the arts because science deals with fact or with information. As Roszak puts it, "When the modern Prometheus searches for knowledge... he brings back...many kinds of information," or "At one end (of our experience) we have the hard, bright lights of science; here we find information."

---

4 Weisskopf, V. F. (1975). The Frontiers and Limits of Science, *Bulletin of the American Academy of Arts and Sciences*, **28**(6), 15–26.

The scientist, as observer, is of course in some sense searching for the facts. I say "in some sense" because in the absence of any theoretical preconception or organization the so-called facts are close to meaningless. The full range of possible human experience is so large, the variety of ways of looking at the same events so diverse, that it is almost impossible to record our experience in any sensible way with no preconception as to what is significant – and of course these preconceptions often are the boldest strokes of organization.

Crucial questions take their meaning only in the context of the conventions and beliefs of the period in which they are formulated – one reason it is so important to teach science in its historic context. The results of the Michelson–Morley experiment, for example, were startling in the light of the theoretical preconceptions that had evolved from Copernicus to Maxwell. Aristotle or Ptolemy would have greeted their result as self-evident, as more support for the well-known fact (?) that the Earth is at rest at the center of the universe.

Even if we could separate the mining of facts from the invention of theory, the view of science as information misses completely what seems to me is the most remarkable and, in a way, the most astonishing achievement of scientific thinking – that is the creation of order, the organization of this so-called information. It is surprising that it can be done at all. As Einstein expressed it: "The most incomprehensible thing about nature is that it is comprehensible."

The relation between our organization of experience and "the true organization" is much debated. Perhaps, as science advances, we do approach more and more closely to what is truly the ordering of the world. I have nothing to add to the deep philosophical argument that engulfs this question. But, in the midst of the process, it is evident that the actual orderings we have created – temporary as they are, they are the only orderings we have created – have been created by humans and are not of themselves in the phenomena. They are human inventions, rather than the result of a mining operation. This is evident because the same "facts" are ordered in what must be called strikingly different fashions from one generation to the next.

The planetary orbits will approximate Keplerian ellipses, whether we follow Newton, Schrödinger, or Einstein. The assumptions, meaning, implications, and interpretation are strikingly different, but the structure (in the domain of planetary orbits) is identical. What is constant in scientific theory are just such structural relations as those between inverse square forces (or their geometrical equivalent) and elliptical orbits. It is one reason that the interpretation, the "meaning" of a scientific theory, comes last – for interpretation can change entirely without modifying structural relations.

Quantum electrodynamics provides us with an interesting and current example. The agreement between observation and calculation for such quantities as the magnetic moment of the electron or the Lamb shift must be regarded as incredible. Yet it is commonly thought that the axioms on which quantum electrodynamics is founded – or at least the methods of calculation – may be inconsistent. It seems very likely that these will be modified in the future. Yet one can say with complete assurance that the structural relations that lead to such agreement between experiment and theory will remain valid (if approximate) forever.

It is when the scientist creates order, it seems to me, that he is closest to the artist. For the scientist, with his own techniques and within his own medium, creates an ordering of the phenomena just as the artist creates an ordering of the phenomena with which he is concerned: the painter, for example, light, color, and form; the composer, sound and so on. This ordering is as much the scientist's personal vision of the world as is the artist's. It may seem that the artist is less bounded by the "facts" or what is "real," but he is circumscribed by what he hears and sees as well as by the conventions, the ways of regarding the world given him by his predecessors. If modern music and painting seem self-consciously detached from conventional forms, it is not clear that they are any more so than the average theoretical speculation published in *Physical Review*. As time passes we will very likely find among these works the conventions with which we will organize the future.

What we take for granted has not always been obvious. Today's physicist lives in a world of wave functions, matrix elements, and non-commuting algebras. A generation ago our elders struggled to understand atomic phenomena using semi-classical analogies. An ordinary magazine will print full-color pages that look vaguely like something Renoir might have brushed to amuse his children. The Impressionists' vocabulary has become part of our dictionary of clichés, with little memory of the derision that greeted their early exhibits. The composer of a sonata we now so admire was mercilessly criticized for writing complex noise. What he first heard in the ear of his mind we now accept as normal – further, we expect music to sound as he created it

It is surely here that we find a closeness of purpose that unites storyteller, painter, musician, and scientist. What each gives us is obviously not literal representation but a personal vision of the order of the world. This vision, when it is most original, often seems most strange. But when we learn to see, hear, or understand, we find that the world comes to appear to us in this new guise.

It is in the nature of the order characteristically employed by the scientist that we appear to come to a crossroads. For traditional painting and poetry have created order in human terms. It is righteous wrath that moves Apollo to fire arrows of pestilence into the Greek camp before the towers of Troy. It is the thirsty spirit that the savage fears in the vertical boiler. The universe feels as we do and acts as irrationally.

Science, rising from the fire of magic and arcane conundrums of numerology, has, of course, completely discarded such anthropomorphic concepts as spirits inside boilers. Not, as might be suggested, because they are unscientific, but rather because they are imprecise and uneconomical. The remarkable world-view we have constructed does not require such assumptions. For science, economy, precision, and internal consistency come first. (It is also true, of course, that today's painters, poets, and musicians have deliberately and quite consciously attempted to cleanse their work of traditional ordering and often challenge their audience and make it as nervous as does science.)

There would be no problem, except that the very success of science in constructing an order that obviously works has made it unfashionable to cling to belief that science has not found necessary to assume. And so cocktail-party conversation that begins in high fashion following the latest "scientific" thinking concludes, "but you have not proven that God does not exist."

Perhaps the most important result is psychological. Medieval man, for example, could live without embarrassment in a world of which he was the center, a world whose reason, awareness, and concern were centered about him, just as the motion of the stars and that of matter, heavy or light, was centered about the Earth; a universe built around the drama of salvation where, as in early Renaissance painting, all things had a purpose – an almost magical luminous world where all of creation from the angels to the beasts, even the inanimate stones, knew their place and their relation to everything else.

Possibly it is too much to say that modern science forces the abandonment of such a world, but its success makes one less comfortable living there. Justifiably or not, it is no longer easy to believe that the world has been constructed about man; that all of creation, no longer centered about the Earth, the result of the motion of particles subject to mechanical laws, is yet directed and ordered with man as the principal character in a grand drama.

The seeming conflict between science and belief is, however, highly exaggerated – the long conflict between science and religion intellectually completely unnecessary. Science never said, "God does not exist." Was it necessary for Christians to believe that the Earth rests immobile at the center of the universe, or that man was or was not descended from the hairy ape? It seems almost accidental that the Church fell into these positions. Was it more than his great respect for Aristotle that led St. Thomas to "reconcile" his thinking with that of the church fathers? Science rarely tells us that we cannot believe. Rather, to paraphrase Laplace, science confides, "We didn't need that hypothesis."

But the source of that widely felt emotional rejection of science, in my opinion, is not so much that science threatens our values; rather it is that science is sometimes felt to remove the basis for the meaning of our own experience. It is as though science, having no place for our humanity, comes between ourselves and our own feelings, thus providing the intellectual basis for that unloved, unwanted, bureaucratic, and computerized world of 1984.

We must, of course, distinguish science as observation and explanation from experience itself. Knowledge and understanding, even understanding of experience in terms of "physical phenomena," is not a substitute for the having of the experience. For experience involves nerve endings as well as the brain – whereas understanding engages the brain alone. This, it seems to me, is neither a limitation on science, nor is it threatening.

I would say that, even in a most extreme and mechanistic guise, and properly understood, science poses no threat to what we should regard as most precious: the value of our individual experience. But science can pose a threat to what we would like to believe. This, it seems to me, is the substance of Roszak's accusation: "I have another monster in mind . . . one who is nobody's child but the scientist's own . . . the monster of meaninglessness . . . The existential void where modern man searches in vain for his soul . . . " For, if we can base our own worth only on some particular structure of belief – if the only meaning we can give to our experience is that which results because the universe knows and cares that we exist – science becomes a threat when it undermines the rational basis for this belief: not because we are denied the possibility of believing, but because certain beliefs cannot reside comfortably in the head that has been taught what science knows.

Existentialist authors have explored with passion the loneliness of man in a universe that does not know he is there. No one will improve on Franz Kafka's *The Trial* and *The Castle* in posing the eternal questions: why are we condemned to die? Why are we denied entry into Paradise? But must we believe that our experience can have meaning only if the universe feels and palpitates as we do?

"I must have had a longish sleep, for, when I woke," says Camus' stranger, in prison and condemned to die, "...the stars were shining down on my face. Sounds of the countryside came faintly in, and the cool night air, veined with smells of earth and salt, fanned my cheeks. The marvelous peace of the sleep bound summer night flooded through me like a tide. Then, just on the edge of daybreak, I heard a steamer's siren. People were starting on a voyage to a world which had ceased to concern me forever..."

And, in what could become the guiding lantern for adult humanity: "...gazing up at the dark sky spangled with its signs and stars, for the first time, the first, I laid my heart open to the benign indifference of the universe. To feel it so like myself, indeed, so brotherly made me realize that I'd been happy, and that I was happy still."

3

Can Science Serve Mankind?

The naïve scientific optimism of the nineteenth century has been replaced by cynicism regarding the ability of science to serve mankind. Some blame science for the breakdown of the social order of an idealized past. Does science serve mankind, or does mankind serve science?

This essay is based on a talk given at the opening ceremony of the conference "Science in the Service of Mankind," Vienna, Austria, July 8–14, 1979.

The scientific optimist who wrote in the 1808 *Elements of Natural Philosophy*: "The great object of science is to ameliorate the condition of man, by adding to the advantages which he naturally possesses,"[1] is no longer with us. He has been replaced by the environmentalist, the conservationist, the consumer advocate, and the professional demonstrator who criticize every aspect of science and most other human activities – who, regarding the splendor of this gathering and the obvious prosperity of its participants, might suggest that the appropriate question is: Can mankind afford to continue to serve science?

As is evident from the topics covered in this conference, in a material sense science has provided, and continues to provide, solutions for many problems. It is obvious that life as we have come

---

1 Webster, J. (1808). *Elements of Natural Philosophy*, Philadelphia: B. and T. Kite, Fry and Kammerer, p. v.

to expect it would not be possible without the material fruits of science. And it is surely true that material problems that remain: problems of overpopulation, energy, disease, poverty, environment among others are enormous; and that there is no way of handling such problems unless new means are found and further, that the world's population does not simply increase to absorb increased wealth due to technological advances leaving us with the same problems. Increasing population strains every new scientific and technological achievement, and necessitates the introduction of highly efficient means of production that usually produce other problems. (It does tend to be forgotten, however, that the level of pollution in London today is far lower than a few hundred years ago.)

With all of the public respect for and admiration of science, many people share, to some small extent, the feeling of science's bitterest critics. Although it is not likely that science alone is responsible for the breakdown of the order we associate with former times (not troubling ourselves with such questions as whether this order was as benign as is suggested in retrospect – a winter scene of Bruegel or *The Temptation of St. Anthony* of Hieronymus Bosch), although the breakdown of this perhaps sentimentally recalled ordered society is more likely a consequence of social changes: migration from the country to the city, industrialization, etc., there was no doubt a contribution due to the increasing dominance of scientific thinking. And in the end the social fabric – not one, but one after another – has been shredded with little presently to replace what has been lost. Our disappointment is perhaps exacerbated by the end of the nineteenth century false hope that somehow scientifically generated values could replace those made increasingly untenable by the scientific world-view.

A consequence for the twentieth century is a society perhaps characterized as a theater of the absurd – not in the cosmic sense but on a very individual level: a society in which there is no general faith or belief to guide action, in which the individual is often so separated from the important social consequences of his actions that individual social action seems meaningless, a society which seems plagued by what might be characterized as a generalized problem of the commons. I do not believe, as is sometimes said, that the average citizen is incapable of rational considered decisions; rather we all have been placed in a situation in which such decisions (beneficial for the individual) are not beneficial for society as a whole.

I think this applies with almost equal validity to irresponsible behavior by Americans in their use of gasoline, by Russians on some collective farms, as well as for peoples of underdeveloped areas who produce (often as protection against old age) too many children. In each case the "socially conscious" individual sees his attempt to behave for the common good mocked and vitiated with statistical certainty, by the behavior of the others.

At this conference we will explore many aspects of science's material services to man and will attempt to pinpoint remaining problems of importance. But it is also possible that from such deliberations we might eventually arrive at a world-view that leads to a rationalization of the social situation. I don't mean at all an Orwellian world but what is perhaps the opposite: a world the way Adam Smith hoped it would be, a world in which behavior is not regulated from above but rather, as in a generalized market place, a world in which the individual can enjoy the consequences of clever and socially worthwhile behavior, a world in which the individual acting rationally in his own interest benefits society as a whole.

I realize that this may be asking too much of a scientific conference or of scientists since no doubt this is more than a purely scientific problem. It is a problem that can sometimes be attacked by appropriate taxation or financial incentives. But perhaps it is not too much to hope that our view of the world that, no doubt, has attributed to the destruction of the "old faith" may some day contribute to a new rationalism in human affairs.

The early-nineteenth-century scientific optimist is no longer with us. But I believe that the choice is not one between optimism and pessimism. One can as easily create sceneries with happy and unhappy endings. The choice, it seems to me, is between those who will try to agree on a path that can lead to success and work to follow it and those handwringers who can only point out the dangers and the risks that line any path, and who will do nothing. For it is obvious that science can serve mankind. Whether it will be enough and quickly enough to deal with the enormous problems that face us, I don't know. One can only hope. And it is sure that unless science does serve mankind there is little hope.

Let snobs scoff at our material achievements. In this city where Mozart died so young of a disease possibly cured today with a few pills, such achievements cannot be regarded as completely worthless. And no matter how bleak the outlook appears, in this country where so few years after the emergence of mankind Mozart was born and shared with us the fruit of his incomparable genius, one cannot be totally pessimistic.

4

Modern Science and Contemporary Discomfort: Metaphor and Reality

Has modern science made our universe "an unbounded theater of the absurd," or is this anxiety misplaced? What do we mean by "the scientific method," and how does it differ from other methods of understanding our world? Is science a description of reality? Is it metaphor or truth? Who decides what science is, and what is to be taught in classrooms?

This essay is based on an essay originally written for The Rights of Memory: Essays on History, Science and American Culture, *edited by Taylor Littleton, and published by University of Alabama Press in 1986. It is based on a lecture given at Auburn University.*

1

Robert Penn Warren has spoken on "the use of the past." One such is surely to situate the present. If for science the present seems unfriendly we might, for comfort, recall the inscription on the dome of the great hall of the National Academy of Sciences in Washington:

> TO SCIENCE
> PILOT OF INDUSTRY
> MULTIPLIER OF THE HARVEST

EXPLORER OF THE UNIVERSE
REVEALER OF NATURE'S LAWS
ETERNAL GUIDE TO TRUTH

The nineteenth-century optimist who wrote these words is, alas, no longer with us. Instead we have the professional demonstrator, the fundamentalist preacher, and the moralist legislator. Perhaps it is too much to say that we have fallen on the winter of our discontent.

But there are cold winds that blow from Washington, and the chill of indifference, even hostility, from the country as a whole, makes it somewhat unlikely that these words would be written today.

Possibly we have all become more sophisticated. We agree that science is useful, if a bit esoteric; scientists are not "with it" – like coiffeurs, fashion designers, or rock stars – nor are they as rich. There has been much noise about the evils of science and technology, but we know quite well that life as we have come to expect it would not be possible without the material fruits of science. And it is to science and technology that we look to solve such vast current problems as those of energy, disease, and poverty (also to solve the problems that result from the solutions), and to provide the means to ensure that the world's population does not simply increase to absorb the increased wealth that results from technological advances, leaving us with the same problems.

All of this appears to me to be more or less obvious. Yet when we are confronted by difficult practical judgments concerning science, an irrational element often dominates. In discussions concerning the application of science to everyday affairs there lurks an uneasiness that is troubling enough to fuel the paranoia that seems so often to surface. I suspect that this is so because even those who express respect for and admiration of science share, to some small extent, the feeling of science's bitterest critics.

Writing recently in *Daedalus*, Theodore Roszak, not an admirer or a lover of science, invokes Mary Shelley, who created the image of the Frankenstein monster. However, he has another monster in mind that troubles him as much as all the others:

one who is nobody's child but the scientist's own and whose taming is no practical task. I mean an invisible demon who works by subtle poison, not upon the flesh and bone, but upon the spirit. I refer to the monster of meaninglessness. The psychic malaise. The existential void where modern man searches in vain for his soul.

Further, he asserts that "it is science which has made our universe an unbounded theater of the absurd."

This image of the Frankenstein monster is persistent and frightening. Are we afraid that science, like Victor Frankenstein, creates monsters, or – what is even worse – that science is itself the monster, destroying our values and our human worth?

There is a striking moment in the Mel Brooks film *Young Frankenstein*. Near the conclusion of this farce, by a sequence of events too complex to detail here, the wordless and frightening monster suddenly speaks, and a remarkable transformation occurs. Though he is unchanged physically, the instant he speaks we are no longer afraid – as if with the utterance of words he ceases to inspire terror.

I think it may be that, for many people, science poses a threat that is perceived and felt, if not understood. Science is powerful and, like Mary Shelley's monster, moves with a directive of its own, oblivious to human value or desire. If there is an object to one more talk on this subject, it is to give words to this monster so that we might regard him without terror.

2

The monster that science is, opposed to those we create (the monster that has made our universe "an unbounded theater of the absurd"), arises more likely not from our material achievements, nor from what harm we have caused, but rather from the world-view we have created (or those we have destroyed). Science has presented us with a truly remarkable vision of the world, a metaphoric view of reality that connects events from distant galaxies to living cells – of which we are justly proud. Let me remind you of some of these wonders: elementary particles and atoms, electricity and chemistry, DNA and living matter. We might conclude that what has not yet been explained we will be able to explain in the future, so that, as Victor Weisskopf has written, "man will eventually understand all of nature scientifically."

Yet it is just this vision that becomes a monster where it shatters or makes difficult to retain other visions we want (or even need) to believe. Still, it is not correct to state that there is a single, timeless scientific view of the world; as science develops, its worldview changes. Between the nineteenth and twentieth centuries, for example, we have gone through revolutionary changes due to the impact of Freud on psychology, Einstein on physics, and Darwin on biology.

Little was dearer to the nineteenth-century physicist than Newtonian mechanics and the wave theory of light. Yet, under the pressure of the new atomic phenomena, the assumptions underlying these theories were abandoned. Causality, a concept that underlies almost all of Western thinking, has been modified. David Hume had suggested that causality was not in the phenomena themselves but was a concept introduced to order our experience – the type of philosophical conjecture that elicits knowing smiles from working scientists. But to bring order to phenomena in the atomic domain, quantum theory – the twentieth-century replacement for Newtonian mechanics – has taken us from a completely deterministic theory to one in which equal causes do not produce equal effects. Einstein redefined time; the comfortable Swiss clock has been replaced in the popular imagination by a Dali grotesque. Possibly most threatening has been Darwin's great concept of evolution. Man, no longer the special – almost private – work of his Creator, has become but one link in an endless evolutionary chain.

Now these changes (even though they don't occur every day) are perfectly acceptable to scientists, but the change in world-views they seem to imply does make some people nervous. Science poses for us, in a particularly acute form, a conflict between what we want to believe and what seems sensible to believe if one follows current fashion. For there are many beliefs that are dear to us, that we wish to retain – even though science does not need them to construct its theories.

A current example is the controversy concerning "special creation." It seems astonishing that such a debate should flare up again – an example of the tenacity with which some views are held. The claim is made that special creation is a valid scientific theory that should be given equal time with the conventional theory of evolution in the teaching of high-school biology. This is so outrageous that it is hard to regard seriously. But such a claim does underline a general misunderstanding of what science is, and raises in a concrete form some interesting and serious questions: abstract questions, such as the relation between metaphor and reality, and very practical questions, such as who is to decide what a subject is and what we should teach in our schools.

One of the uses of the past is to remind us that all is not new. This is not the first time science has been told what it must believe and what it must teach. In a moving chapter ("The Starry Messenger" in *The Ascent of Man*) that some of you may have read, Jacob Bronowski writes of the confrontation between Galileo and the Church. In 1616 the Church ruled that certain propositions were to be forbidden: that the Sun is immovable at the center of the heaven; that the Earth is not at the center of the heaven, and is not immovable, but moves by a double motion.[1]

It has always astonished me that the Roman Church became so committed to a view of the universe that seems so peripheral to the central issues of Christian faith. (Previous errors don't seem

1 Bronowski, J. (1973). *The Ascent of Man*, Boston: Little, Brown, p. 207.

to temper its strong views on current affairs today.) As is often so in human affairs, the path was complicated; in some ways it was almost an accident that the Ptolemaic view of the universe became Church doctrine.

When Aristotle was reintroduced into Europe via Latin translations of newly discovered Arabic translations of his notes, his views were regarded with a good deal of suspicion. There was a time, for example, when the teaching of Aristotelian physics was prohibited in Paris. However, through the monumental efforts of thirteenth-century scholars such as St. Thomas Aquinas (moved, I believe, in great part by their enormous respect for Aristotle's intellect), Aristotle's hitherto suspect views were "reconciled" with early Christian doctrine.

Although the reconciliation was never complete, from this time the Aristotelian conception of the universe became a part of the Christian drama of salvation. Thereafter, an attack on Aristotle became an attack on the Church itself. For such a variety of complicated reasons, by Galileo's time one view of a seemingly straightforward question about the world had become official Church doctrine. The issue was from then on political as much as scientific.

Galileo very likely made grave political blunders. We can imagine a revisionist historian earning his Ph.D. by casting him in the role of villain. Yet we are not profuse in our admiration of neighboring societies that make political issues of scientific questions (Soviet genetics or Nazi physics, for example). I think it must be a matter of eternal embarrassment that the Holy Church went to the length of fabricating, and possibly forging, documents (so that, as you may have heard, the case has been reopened) to show Galileo guilty.

With the threat of torture, it wrung from this great man – seventy years old at the time – an abjuration of the heresy "of having held and believed that the sun is the center of the world and immovable, and that the earth is not the center and moves..." "I abjure, curse and detest the aforesaid errors and heresies," Galileo swore.[2] (But tradition, not willing to accept such ignominy, has it that under his breath he muttered, "Yet it still moves.")

3

It is a common misconception that science is separated from the arts because science deals with fact or with information. As a critic puts it, "When the modern Prometheus searches for knowledge ... he brings back ... many candles of information," or "At one end (of our experience) we have the hard, bright lights of science; here we find information."[3] The scientist, as observer, is, of course, in some sense searching for the facts. I say "in some sense" because, in the absence of any theoretical preconception or organization, the so-called facts are close to meaningless. The full range of possible human experience is so large, the variety of ways of looking at the same events so diverse, that it is almost impossible to record our experience in any sensible way with no preconception as to what is significant. And, of course, these preconceptions often are the boldest strokes of organization.

Even if we could separate the mining of facts from the invention of theory, the view of science as information misses completely what seems to me is the most remarkable and, in a way the most astonishing achievement of scientific thinking – that is, the creation of order, the organization of this so-called information. It is surprising that it can be done at all. As Einstein expressed it, "The most incomprehensible thing about nature is that it is comprehensible."

There is a sense in which the scientist, like the painter, poet, or novelist, is imposing an explicit ordering, created in his mind, on a more or less recalcitrant nature. The scientist's conception of

2 Ibid., p. 216.
3 Roszak, T. (1974). The Monster and the Titan: Science, Knowledge, and Gnosis, *Daedalus*, Summer 1974, 17–32.

Fig. 4.1 *The Ninth Circle of Hell*, by Gustave Doré.

the world is, of course, different in purpose and certainly different in detail from that of the poet. The ordering of Dante may seem imposed on reluctant reality but in the perspective of today's quantum world, mightn't we regard the mechanistic Newtonian ordering, as seen by Laplace, for example, in the same way?

Having mentioned Dante, I can't resist showing you just one of Gustave Doré's illustrations of his remarkable journey, the ninth circle of Hell: three-faced Satan, frozen hip deep in a lake of ice, where, embedded for eternity to a depth appropriate to their crimes, are some of Earth's greatest sinners. On the right, if you look closely, you will see two figures: Dante and his personal guide, the Latin poet Virgil.

Dante was, of course, primarily interested in a moral order of the universe. But it is fascinating to observe how closely he followed the received astronomy and physics of his time. The center of Earth was the center of the Aristotelian universe, and it is the center (and the bottom) of Dante's as well: that point to which all heavy (sinful, earthy) matter is attracted. It is there that the creator of sin, Satan, is forever transfixed. Shortly, Dante and Virgil will begin their difficult climb down Satan's body (clinging to his hairy shanks), to descend to the very bottom of Hell (the center of the universe) and, having passed this point, begin the equally difficult ascent to Purgatory.

> From shag to shag he now went slowly down.
> Between the matted hair and crusts of ice.
> When we had reached that point just where the thigh
> Doth turn upon the thickness of the haunch.
> My leader, with fatigue and labored breath
> Brought 'round his head to where his legs had been.
> And grasped the hair like one who clambers up,
> So that I thought our way lay back to hell.[4]

In a Newtonian universe, their experience would have been more like that of an astronaut: they would have floated. (At the center of Earth, the gravitational force goes to zero.) This only reported experience seems to favor Aristotle.

For Dante, one senses that this invisible moral ordering of the universe is real: good rewarded, enshrined with the weightless and ethereal in God's empyrean; sin and evil fallen with the dross and heavy to the antipode, the center, their closeness to Satan determined by the weight of their sin – in strict correspondence with the physical universe as he believed it to be. It is as real as, and is what gives meaning to, daily experience of treachery, disease, and death.

And two millennia before Dante, in the First Book of Moses, called Genesis, the unknown poet wrote:

---

4 Alighieri, D. (ca. 1321). *The Divine Comedy*, transl. Lawrence Grant White, New York: Pantheon Books, 1948, Canto 34.

In the beginning God created the heaven and the earth. And the earth was without form and void; and darkness was upon the face of the deep. And the Spirit of God moved upon the face of the waters ...

Now it would seem that science is far removed from such metaphor – and in a sense it is. But when we ask what science means, the answer at any time may be regarded at least in part as metaphor. For though the structure of science often remains relatively unchanged, its view of the world changes radically. Is truth Newtonian determinism or quantum chance? Are these really and literally true, or at least part metaphor? If part metaphor, how are they to be distinguished from poetic metaphor? How can we distinguish Newton from Dante, Darwin from Genesis? And which is to be called science?

4

In their attempts to rebut "creationist" arguments, scientists and educators, perhaps in exasperation, have called creationism unscientific. A recent AAAS (American Association for the Advancement of Science) Resolution on Creationism resolves, "that because 'Creationist Science' has no scientific validity it should not be taught as science." A recent article in *Physics Today* echoes this theme: "While theories in science are falsifiable, creationist beliefs are not."

There is a sense in which these statements are appropriate. If the creationist's view is that there was a single creation, it must account, among other things, for the fossil records – and these seem to indicate fairly unambiguously (as my colleague Kenneth R. Miller has pointed out) that there was either continual change or many separate creations. If the creationist's argument chooses to remain vague or confused on this point, it is not science, because it is not well defined (what is said does not depend unambiguously on what was said before).

However, it requires no great effort to construct a creationist argument that is well defined and is consistent with everything we observe: We need only assume that in the beginning God created the heaven and the earth and the fossil records.[5] (For that matter, we can assume that He created everything, including our memory of past events just before I began to speak.)

The problem is not that the creationist view is unscientific, but that in any of its current variations it is not very interesting science.

But there is also a clear and major difference in motivation. Although science has from time to time maintained a world-view (e.g., particles moving in a void as the basis for all experience), the view is second in importance to the precision of the assumptions, the logic of the structure, and its detailed correspondence with the "real world." As Galileo said, we must "make what is said depend on what is said before." His teachers of mathematics taught him this method. It is the elegance and naturalness of the hypothesis and the resulting structure that characterizes what we call a beautiful theory. Science may present a vision of the world; but this scientific vision is the end result of a complex construction whose acceptance depends upon its detailed agreement with experience. The construction – the structure – has more permanence than its interpretation or meaning. This is evident, since from time to time (between the nineteenth and twentieth centuries, for example) the scientific world-view has changed radically. Had we been led to the hypothesis that God created the world in six days and rested on the seventh, this would be our world-view.

For the poet, it is richness of metaphor, ambiguity, and evocative quality of language that are of primary importance. He often begins with a world-view, a moral order, a sense of the meaning of experience (in the case of Genesis, perhaps an attempt to end a meandering, interminable dispute: Where do we come from? How

---

5 Such a view actually was presented in the nineteenth century by Philip Gosse in *Omphalos*, published in 1857 (London: John Van Voorst). This was recently discussed by Stephen Jay Gould in *Natural History* magazine.

did the world begin?), and searches for a means to express this view concretely. But whether Dante's vision is an accurate description of the world as it actually is or whether Genesis is a literal account of creation is almost completely irrelevant. The poetic message is clearly independent of literal truth or correspondence with the "real" world.

In contrast, for the creationist it is the world-view that is a fundamental – in this case, the powerful metaphoric vision of Genesis. He shapes his hypothesis (if he is honest) or the facts (if he is not) to make the structure (such as it is) that evolves consistent with what he is willing to agree is observed, in such a way as to maintain the literal truth of his initial world-view. His object is neither beauty nor power of expression, economy, and consistency of structure – nor even agreement with experience – but the preservation of his world-view. His fundamental problem is not necessarily that he is incorrect or "unscientific," but that he had unfortunately converted magnificent metaphor into trivial science.

If the creationist argument can be made "scientific," how do we deal with the creationists' demand that they be afforded equal time in the schools to "answer" evolutionist arguments? This is non-trivial. For at heart are such questions as: Who decides what science is? Who decides what is to be taught in our schools?

We note first that the so-called creationist theory is not a unique alternative to evolution. There are many creation myths. (In the first two chapters of Genesis there are two.) Which creation myth do we teach as an alternative to evolution? For every view taught in the schools there are hundreds of others – must we give all equal time?

The problem for the school system is whether it wishes to teach science or world-views. If the choice is to teach science, a commonsense and conservative position would be that it is scientists who must determine what is good or bad science. We have found it sound practice to allow those in a profession to decide among themselves what the best opinions and practices of that profession are at any time. Clearly, the profession is not always correct (opinions, after all, change). But if the issue is how best to handle an airplane, how to do open-heart surgery, or what is a legitimate proof of a theorem, we generally let the pilots/engineers, surgeons, or mathematicians decide. Otherwise, we are faced with the spectacle of passionate orators arguing difficult and technical matters before audiences that do not have the competence to choose among the arguments.

Schoolchildren have the right to the best education possible. But it is not they who should decide highly technical differences between the experts. Practical problems can arise: Are biologists so closed to new or different views that rump groups must be set up to challenge them? Such a thing could happen. This is an empirical question and it would seem to me demonstrably false. I know many biologists. They display qualities common to the rest of us. Some are difficult, arrogant, or blunt; others are subtle, brilliant, and flexible. The profession as a whole is constantly debating new ideas. If "creation science" had any serious content, it would be debated just as earnestly as any of a dozen other ideas. What the community is telling us is that this so-called science is so bad or trivial it is not worth wasting time discussing. Yes, they might be wrong. But it is they who must be convinced. If "special creation" is a scientific theory that can rival evolution, this must be argued before those who know the subject – not before schoolchildren.[6]

Such arguments often have no logical resolution. Consider the following table, listing the ionic content of concentrations of axoplasm, blood, and seawater. I am led strongly to the conclusion that blood (largely seawater with red cells added for color) came into being when seawater was first employed by groups of swimming cells to provide a means for internal transportation of various goods and services – an argument, seemingly, for evolution. But, a

6 A somewhat different point of view is presented in Chapter 9, "Is Evolution A Theory? A Modest Proposal."

Table 4.1 Concentration (mM)

Ion	Axoplasm	Blood	Seawater
Potassium	400	20	10
Sodium	50	440	460
Chloride	40–150	560	540
Calcium	0.3×10^{-3}	10	10

The precise value of ionized intracellular calcium is not known.
Source: Data from Hodgkin (1964) and Baker, Hodgkin, and Ridgway (1971).[7]

counter-argument might go; this is merely another manifestation of the economy practiced by an efficient Creator.

There is a lesson here for the teaching of science: science must be taught with the emphasis on logic and structure, rather than as a collection of "facts" that deliver the current scientific world-view. When science is taught as a set of truths, rather than as a process of discovery, invention, and construction, we distort what science is and fall into a trap that makes us vulnerable to those with opposing world-views.

It is the creationists' misfortune to have made their faith dependent on a literal interpretation of Genesis, thus running afoul of science – the supreme specialist in constructing a consistent ordering of the world that is. They are not the first to have maneuvered themselves into such an uncomfortable position. To defend their faith, they have made a bold and somewhat unconventional move: proclaiming themselves a science. By itself, this might be regarded as no more than an expression of desire – the triumph of desire over reason. Perhaps the desire is comprehensible. But the situation has become complicated by moralistic or opportunistic legislators who attempt to legislate desire as reality: to dictate what should be

7 Kuffler, S. W. and Nicholls J. G. (1976). *From Neuron to Brain*, Sunderland, MA: Sinauer Associates, p. 91.

taught in our schools or to choose for us among various possible scientific hypotheses.

It requires a totalitarian mentality – or great naïveté – to legislate such things as the teaching of the creationist view, the moment of beginning of life, or even the value of π. One can legislate that $\pi = 3$ (as has been done), but that hardly changes the ratio between the circumference of a circle and its radius in a Euclidean space. (My own preference would be $2\pi = 1$.)

It is unnecessary confrontation. Science has discarded concepts such as those espoused by creationists – not, as might be suggested, because they are unscientific but because they are imprecise and uneconomical. The remarkable world-view we have constructed does not require them. Perhaps the most important result is psychological. Medieval man could live without embarrassment in a world whose reason, awareness, and concern were centered about him, just as the motion of the stars and that of matter, heavy or light, was centered about the Earth. It was a universe built around the drama of salvation, where, as in early Renaissance painting, all things had a purpose – an almost magical, luminous world where all of creation, from the angels to the beasts and even inanimate stones, knew their place and their relation to everything else. He could believe, if he wished, that the words written in Genesis were literally true.

Possibly it is too much to say that modern science forces the abandonment of such a world, but its success makes one less comfortable living there. Justifiably or not, it is no longer easy to believe that the world has been constructed about man – that all of the creation, no longer centered about the Earth, the result of the motion of particles subject to physical laws – is yet directed and ordered with man as the principal character in a grand drama. And that the Creator did his great work in six days, resting on the seventh.

And it is this, perhaps, that has led to some of the modern discomfort with science – not only among biblical fundamentalists but, more generally, as expressed by such critics as Roszak.

Although it is not likely that science alone is responsible for the breakdown of the order we associate with former times (not troubling ourselves with such questions as whether this order was as benign as is suggested in retrospect – a winter scene of Bruegel or *The Temptation of St. Anthony* of Hieronymus Bosch), although the breakdown of this perhaps sentimentally recalled ordered society is more likely a consequence of social changes (migration from the country to the city, industrialization, etc.), there was no doubt a contribution due to the increasing dominance of scientific thinking. And in the end the social fabric – not one thread only, but one thread after another – has been shredded, with little to replace what has been lost. Our disappointment is perhaps exacerbated by the end of the nineteenth-century false hope that, somehow, scientifically generated values could replace those made increasingly untenable by the scientific world-view.

If we must create a new world-view as well as new values, hopefully this can be done with respect for scientific evidence as well our treasured heritage.

We can earn our place in the past, which we shall soon become, by creating the future. That, concludes Robert Penn Warren, "is the promise the past makes to us." In that promise there is no greater gift than our heritage of metaphor and structure – no greater burden than the heavy weight of outworn ideology.

5

Faith and Science

Why do faith and science often seem to be in conflict? Is such conflict necessary? Past confrontations such as those involving Galileo are now acknowledged as unnecessary. Will current confrontations also seem unnecessary 350 years from now?

This essay is based on a talk given in 1983 on the occasion of the discourse of His Holiness Pope John Paul II for the Inauguration of the Symposium organized for the 350th Anniversary of the publication of the book by Galileo Galilei entitled Dialogo sopra i due massimi sistemi del mondo.

It is gratifying to hear the courageous and thoughtful discourse of His Holiness for the inauguration of this symposium organized on the occasion of the 350th anniversary of the publication of the book by Galileo Galilei entitled *Dialogo sopra i due massimi sistemi del mondo*.

To address ourselves to those issues producing unnecessary confrontation between science and faith in the spirit so movingly expressed by John Paul, to follow the wish of Saint Robert Bellarmine "that useless tensions and harmful rigidities between faith and science be avoided," we must turn our attention to current propositions that produce controversy. Today these are often biological: such questions as the origin of intellect, the nature of consciousness, the meaning of soul, the construction from ordinary materials of the special entity that is ourselves and the definition of life itself.

Let us consider a specific example. It has been shown that the nuclei of adult frog skin cells (among the most highly specialized types an organism possesses) contain the full genetic information so that when transplanted into enucleated egg cells, produce a clone of swimming tadpoles.[1] Might we say that life has already begun in every one of these skin cells – so carelessly sloughed away by the millions every day? Or should we insist that it is a signal provided by the enucleated egg cell that is to be designated life's origin?

A proper scientific answer to this question is not yet available; but it seems likely that the best answer will be that the question of when life begins is no more meaningful than the raging controversy over those once forbidden propositions:

that the sun is immovable at the center of the heaven; that the earth is not at the center of the heaven, and is not immovable, but moves by a double motion.

The chamber in which Galileo was tried is now part of the Post Office of Rome – responsible for the speedy delivery of Italian mail; and there is no longer controversy over the questions for which he was disciplined. To properly draw the lessons from the past we must apply them to the present. I trust that in that "frank and open dialogue" between scientists and Church leaders urged yesterday by John Paul, the appropriate conclusions will be drawn: it is the gravest error to make faith dependent on a view of the world derived from "a culturally influenced reading of the Bible." This is true for biological as well as cosmological, recent as well as ancient questions. For such a path leads faith to just that confrontation with science we now realize is damaging and unnecessary. I can think of no better commemoration of Galileo Galilei, one of humanity's greatest intellects, and no better fulfillment of John Paul's memorable words than to avoid committing this error again.

1 Gurdon, J. B., Laskey, R. A., and Reeves, O. R. (1975). The Developmental Capacity of Nuclei Transplanted from Keratinized Skin Cells of Adult Frogs, *Journal of Embryology and Experimental Morphology*, **34**, 93–112.

6

Art and Science

Why is art often thought of in opposition to science? We do both a disservice by focusing on their differences while neglecting their similarities. Is it possible that at a deep level these two have similar goals?

This essay is based on an article originally published in the journal Daedalus, *1986, vol. 115(3).*

We might ask why this question of the difference between art and science is posed. Why are we not talking about the similarities or differences between a painter and a musician, or between a sculptor and a poet? Why is the distinction between the scientist and the artist of special concern? I would like to suggest that this is less an intellectual than a sociological question. It is perhaps because, unfortunately, more and more people who call themselves scientists live, think, and work in a different part of society, both intellectually and socially, from the part inhabited by those who call themselves artists.

There are indeed enormous differences of style in science. Yet, one characteristic of science is that the way things are said in science is almost always less important than what is said, whereas the way a concept is expressed in art is almost always more important than what is expressed.

There is a vast difference between science being created and science being presented (in textbooks, for example). For science being created, I see a closeness between art and science. Creating a painting or a sculpture, a piece of music or a piece of science, involves processes in which intuition is important. The process is not entirely logical – quite distinct from formal presentations in textbooks.

An excellent example is Galileo's treatment of falling bodies. What is important is not that he was right and that Aristotle was wrong, but that he chose to ignore what experience tells us, that a heavy stone falls faster than a light one. Galileo held the view that this small difference is not important. He chose to regard as relevant what would occur if air resistance was removed. Now the decision that the air is extraneous represents a point of view, a new way of looking at the phenomenon. It is a tremendously fruitful approach, for the inclusion of the resistance of the medium as part of the fundamental phenomenon was one of the reasons the Aristotelian approach had come to a dead end. In this respect, the scientist and the artist are in very deep accord because there is a sense in which the artist selects; he decides what to look at, what to emphasize; in a certain sense, he presents a world-view.

Consider the two different ways of looking at the world represented, for example, by the Impressionists, and the Cubists. Presumably, both live in the same world, but a decision is made to select particular aspects of it, to organize it in a certain way. A painting of Monet could be regarded as Monet's point of view of what to look for, or how to regard experience. I think there is clear evidence for this. If we had had *Time* magazine in the early nineteenth century, it would have been somewhat unlikely to find a picture that looked like a Monet. On the other hand, in recent years magazines, popular or otherwise, contain sketches that may or may not have the same skill or originality, but clearly display a way of representing the world that was created by the Impressionists or other recent movements.

I would like to give what I regard as a magnificent example of creativity in science, to present some of the process, and some of the ambiguities and problems. Consider Einstein's Theory of

Gravitation. Einstein was fascinated by what fascinated Galileo: that (neglecting air resistance) all bodies fall to earth at the same rate. Now, in Newton's theory, this is explained as a consequence of accidental coincidence of two very different constants. Einstein wanted to produce a theory in which this phenomenon was really the heart of the matter. He worked on the problem for ten years, and produced the famous Theory of Gravitation in which the gravitational phenomena were explained essentially in a geometrical manner. It is intriguing that for a vast range of phenomena (essentially everything we encounter in our ordinary daily experience) this theory is practically identical in its consequences with the Newtonian theory that came before.

One can imagine, by extension, Einstein producing a geometrical theory that had a completely different set of assumptions, a completely different world-view, but that gave results that were identical in every respect to those of Newton's theory. Which then would be true? We would have an example (which I admit does not occur very often) of two theories, identical in their consequences in the observable world, but whose underlying assumptions – points of view – are completely different.

I would like to pursue the question raised earlier: why can we no longer paint in the Renaissance manner? We presume that a painter today is at least somewhat conscious of what has gone on in the last few hundred years. If he tried to paint in the Renaissance manner, his work would seem inauthentic. While elements of Renaissance style can be repeated in an ironic sense today, to do it straight would be peculiar. It would be as though the artist were not aware of the interim period. The painting might be technically excellent, but it would be anachronous.

There is an analogue to this in science. All physicists study electrostatics. However, we are not likely to publish papers solving problems in electrostatics today. But with the occurrence of a new theory that requires the old technique, the old technique is revived; it becomes current, and very modern. The issue is not that you cannot use old techniques, it is not that you cannot paint and make a reference to what happened before, but that it somehow has to be done in the context of what is considered important today, and what has gone on in the interim.

But why are we talking about art and science? We are looking very hard for areas common to art and science, between artist and scientist. They certainly exist. Both are human beings, both create things, although their products are different. Why are we searching for these common areas? Why do people claim that it is going to help a scientist if he is familiar with art history, or if he knows how to look at a painting? I think we would all agree that it will make him a richer human being. But it is hard to believe that it is going to help him be a better scientist. The same is true for an artist. It would be wonderful if artists knew something about science; it would help them as human beings. But the chances that it will help them in their art are, I would say, negligible. They might as well learn about banking, or international trade. Perhaps one reason is that, if you look at our civilization, and you consider the greatest things that our civilization has achieved, you would certainly list the creations of art, the creations of music, and the creations of science. I think one might also note as a fact of the sociology of our current society, that artists and scientists not only interact with each other very little, they do not even talk the same language; in fact they are suspicious of one another. There is a common fiction that scientists are inhuman; the stereotype of the artist is to be engrossed in "feeling." The creators of these magnificent manifestations of our culture do not talk to each other; rather, they have become hostile. Scientists in general have some interest in music, and occasionally go to a museum. Artists, along with other nonscientists, may occasionally watch *Nova* or peruse *Scientific American*, but they have almost completely separated themselves from any kind of scientific thinking, any of the language of science, the mathematics, any of the structure of science. The schism is a profound cultural concern. How is it that these superb creations of our civilization have reached a point where there is so little interaction between them? In a book I wrote attempting to present physics to non-scientists, I tried as hard as I could to avoid equations. Critics told me, "The problem is you mention Shakespeare and Dante as though the students should know who these people are." I replied that I had thought I should try to skip the mathematics, but never thought I should omit any mention of our cultural heritage. Has the level of college education in this country dropped so low that not only mathematics, but any serious and disciplined intellectual approach is no longer acceptable?

In business, the criterion of success is extremely unambiguous – you either make money or you fail to. In science, success is not quite so easily defined, but it is still fairly well accepted – and colleagues recognize successful scientific work. Maybe one of the problems in the world of the arts is that such unambiguous criteria do not exist, or at least they are not commonly accepted even among artists. Some criteria do exist, some general acceptance is there, but if you consider the contemporary scene – at least from the point of view of someone who has to look in from the outside – you have the feeling that even among artists there is no general consensus.

I take exception to the idea that science is more likely to become a tool of the government than any other endeavor. This is not a problem of science, but a problem of politics. An authoritarian government will make the arts, the means of expression, subservient to the state, with devastating results. I would agree that we have to be sensitive and aware at all times of abuse of power of the state, but it is not a problem that is peculiar to science.

One problem we do have is that big science depends on big money, and big money comes from the government. There is still small science, and perhaps small science does the best work, for which it is fortunately not totally dependent on the government. We had better keep it that way.

7

Fraud in Science

Fraud is a problem in science, as it is in many human activities.
Instances of fraud are relatively rare, but when they happen they tend
to garner a lot of attention. Does fraud undermine the sanctity of
science – and by the way who ever said science should be sanctified?

This essay is based on an article originally published in the George
Street Journal, *16(15), in 1991.*

Recent highly publicized cases of "scientific fraud" have elicited much excited comment. We hear cries for watchdogs and oversight committees. We read perturbed editorials speaking of a scientific Watergate. But let's put the situation in perspective and avoid creating something worse than the disease – a cure.

In science, as elsewhere, lying, cheating, stealing, and fraud are distinctly unpleasant. Because scientists are human, one must anticipate that in their daily as well as their scientific activities, they will occasionally stray. What is surprising is our surprise. I recall my mother pointing out to me, a hopeful teenage scientist, an embarrassing situation in which scientists were involved. My smart-aleck response was, "So what? Everybody does it." "But," she responded, "these are scientists."

Well, the fact is that scientists do it, too. Under pressure for promotions, grants and the necessity to publish, we can be "reasonably assured" that corners are cut, work is sometimes sloppy and, occasionally, I suppose, there is deliberate fraud.

The idea that science is a special enterprise different from other enterprises, that scientists engage in a mysterious method separated from other human methods, is a myth that no amount of explanation seems able to overcome. The scientific enterprise, in fact, is very simple. It is to discover what the world is: to separate the way things are from the way things might be, then to understand how everything that is ties together.

Few things are more damaging to this enterprise than an incorrectly reported fact. It misleads us in fabricating the structures we struggle to create. I have personally experienced situations in which we have been misled for years, even decades, by reported experiments that turned out to be in error.

So why am I so little incensed by scientific fraud? The answer is simple. A working scientist lives in an ocean of uncertainty. Hypotheses are sometimes foolish. Experiments are sometimes sloppy and hastily done. Reported results may or may not indicate what the experimenter believes they do. Fraud is the least of our troubles.

We are aware that many experiments are extraordinarily difficult to do and the best and most honest can mislead. In addition, as we have seen in recent instances involving high-temperature superconductivity and cold fusion, the normal human desire to appear in the papers, to announce results first, to hold the first press conference, can lead to sloppy work. I recall the amusement and skepticism my colleagues showed for some of the early work in cold fusion, work in which individuals all over the world employed delicate and hard-to-use instruments with which they had little experience and within days reported results to the newspapers in hastily called press conferences only to retract these results a week or two later.

How, one might ask, can we survive in such a situation? The answer is not difficult; it is the daily working technique of almost all of us. One stands back a bit, waits until the dust settles, waits until the difficult experiments are done. One doesn't rush about trying to explain every bizarre result that surfaces. As Francis Crick once said "It's most important for the scientist to know what and what not to believe."

The reason that fraud in science is such a minor problem is that to indulge in it is self-defeating. A fraudulent result will not be duplicated by other laboratories, so any gain will at best be temporary. A few such instances and the scientist is branded something worse than a fraud. He is branded as a person whose results are not reliable. In the end, his papers are not taken seriously or are not read. For a scientist, there is no worse fate.

Thus, the problem is self-correcting. The simplest way to deal with a potential instance of fraud is not to spend hundreds of hours verifying whether the data were real or fake, but simply to attempt to duplicate the result. If no one can duplicate it, it's not there.

If we go so far as creating a bureaucracy in universities, agencies and/or (worst possible case) Congress to look over the shoulders of scientists, we will have created another monster that will add to the many that already are making us too inefficient, too slow, and too expensive.

Science in America is not as healthy as it once was, nor as it should be.[1] If we wish to improve the health of science, increase the quality of work, eliminate shortcuts, the rush to publication, the hastily called press conferences, we might investigate what it is that pressures scientists. We might look seriously at the difficulties that face young scientists, at the bureaucracy that encumbers the established, at unnecessary expense due to too much administrative regulation. We might also look into the legal situation involving intellectual property rights that drives people more and more to unwanted secrecy to protect their results. Science has many problems, but fraud, sensational and titillating as it is, is not one of them.

---

1 Its health has not improved since this article was written.

This is not the time to worry about scientific Watergates. Rather, we should worry about the slow strangulation of this great enterprise – one of the intellectual wonders of the world and a driving force for our future economic well being.

8

Why Study Science?
The Keys to the Cathedral

It is often said that it is important for students to study science. Is it really necessary for those who don't plan on careers in the sciences to study or understand science? Who should study science and why?

This essay is based on an article originally published in the Brown Alumni Magazine *in July 1990.*

On this hundredth anniversary of Kamerlingh Onnes' discovery of superconductivity, we may well reflect on the technological and theoretical marvels that are the descendants of his discovery. We may also reflect on the benefits science has showered on us (antibiotics and modern electronics to mention just two). Young people who want to participate in this adventure on a professional level must become technically proficient, so of course they must study science.

But why study science if you don't intend to become a scientist? Why should a lawyer, businessman, or artist study science? We might just as well ask: Why should they read Kemal, Kaya, or even Shakespeare? Why should they watch TV or drink beer?

I would like to present the antique and possibly quixotic view that we should study science because it can give us pleasure. Now the notion of pleasure associated with the physics or chemistry we remember from high school is a hard sell. This, I would say, is one of our great current misfortunes.

For, to my mind, science, in addition to all of its practical uses (and I assure you, we are not rich enough to ignore those practical uses), when properly viewed, presents a vista of stunning proportions that along with music and literature ranks among the great achievements of our civilization – not only because of its intellectual grandeur, but because of the pleasure the awareness of its remarkable and stunning proportions can give. I note, with sadness, that this vista is seemingly permanently closed to most who live on this planet.

In what must have been adolescent enthusiasm, I wrote a book intended to bring physics to non-scientists. I did so, if I may paraphrase, so that the non-technical person could enter the cathedral of physics "to touch the stones themselves."

However, it turns out not to be possible to enter the cathedral without a certain amount of technique. Just as we cannot appreciate literature without first having learned to read, or play a Mozart sonata without first having done finger exercises, we cannot grasp the beauty of science without acquiring a minimum level of technique. We live in an era of instant gratification, but life's deepest and most satisfying pleasures take longer to enjoy.

The problem is to capture in a reasonable amount of time the structure, meaning, and beauty of science in an honest way, without overwhelming the non-technical person with detail.

Anyone who has tried to do this knows how difficult it is. Yet many universities offer such courses; those that do not should be encouraged to do so. These courses are typically put together with great care by scientists who are leaders in their fields, expressly designed to convey to the non-technical person some of the meaning and beauty of their subject. In my opinion, this is as essential a part of every civilized person's education as learning to read or write, and is surely a vital part of the education of every student.

Why study science? One can live without understanding science, just as one can live without song, dance, painting, or literature; but who would want to?

9

Is Evolution a Theory?
A Modest Proposal

Critics of Darwinian evolution assert that it is "only a theory." But what is a theory? And is intelligent design a theory that is a credible alternative to Darwinian theory? Why do we refuse to discuss the ongoing debate in the science classroom?

Since I am a card-carrying member of the scientific establishment, and might possibly be accepted as one of the "Eastern Intellectual Elite," my position on the current, somewhat astonishing "debate" on the teaching of evolution should be as predictable as the orbit of Mars calculated according to Newtonian physics. And it probably is. But today, we are quantum physicists; there may be a degree of uncertainty or unpredictability in our positions.

If Darwinian evolution is not a theory what could it be? It was proposed by Darwin as an explanation for an astonishing variety of facts concerning the interrelatedness of the many species now existent as well as the origins of these species and the evolution of life on Earth. It explains anatomical progressions, DNA, protein and other chemical similarities between different species.

It is one of the more remarkable theoretical structures created by the human mind; Darwin's evolution stands with the great theories of the physical world: remarkable structures that explain and connect vast varieties of phenomena – from the behavior of electrons at the lowest temperatures to the origin of the matter in the fiery instants after the creation of our present universe in what is called the Big Bang.

And as is the case with scientific theories explaining nature as well as detectives' theories solving crimes, they are rarely perfect or complete. There are often gaps, possible inconsistencies, and sometimes seemingly contradictory evidence. In addition there are usually possible alternative explanations.

When a theory is accepted, it is because it appears to be the best, most complete account of the preponderance of evidence available.

Part of the current controversy concerning the teaching of evolution in our school systems concerns the presentation of other possible explanations (creationism, intelligent design, etc.) of the same evidence. None of these is very fashionable or attracts much attention in the scientific community. Though our lack of interest is no doubt justified, I would like to suggest that we are too rigid in our response, and are missing a remarkable opportunity to convey to young students what science really is.

One of the worst failures in the teaching of science is its presentation as a collection of facts, a presentation that misses the extraordinary and dramatic interplay between observation – to distinguish the world as it is from all the ways it might be or we might wish it to be – and explanation. It misses the passionate, emotional, and often personal arguments in the debates between proponents of different explanations that contrast remarkably with the dry and seemingly fixed in concrete textbook recitals. Great unifying ideas that took centuries or even millennia to formulate are presented as colorless – even obvious – facts as though they were mined from the earth rather than invented by some of the greatest minds that have inhabited this planet.

What an opportunity then to present science as it really is at its best: present the evidence and the proposed explanations – those favored by our community and those not. A substantive issue is how much class time to devote to various ideas.

I propose something along the following lines.

Most of class time devoted to presentation and discussion of the facts – the fossil record, carbon dating, common DNA, genes and proteins in living creatures, anatomical and chemical similarities, etc., the basis of various estimates of the age of the Earth as well as the galaxy and our universe, followed by a discussion of proposed explanations: Darwinian evolution, intelligent design all the way to God created the heavens and earth with all the fossil records some 4,000 years ago (or with all our memories 10 minutes ago).

One should also present problems with each explanation – supposed gaps in the fossil records, or if the design is intelligent why do we suffer from lower back pains (do we really live in the best of all possible worlds or the best world possible?). Whether or not this would result in an adequate presentation of Darwin's theory, it might lead to a more mature discussion of how science functions.

Perhaps we could pull back from the present confrontational mode in which the teaching of evolution has become so charged with emotion that, we are told, many high-school teachers (and perhaps some university professors) simply avoid the subject. If we could all lighten up a bit perhaps we could have some fun in the classroom discussing the evidence and the possible proposed explanations – just as we do at scientific conferences. I would leave it to our students to decide what is the best, most consistent, most sensible explanation. My guess is that they are more astute than their elders.[1]

---

1 This is not entirely the point of view expressed in Chapter 4, "Modern Science and Contemporary Discomfort: Metaphor and Reality." However, they are not totally inconsistent.

10

The Silence of the Second

The Second Amendment is silent on the issue of whether the right to keep and bear arms is absolute. Does it guarantee the right to carry loaded weapons into a kindergarten classroom?

In his famous parable, "The Silence of the Sirens," Franz Kafka tells us that the sirens have a still more fatal weapon than their song, namely their silence. "It is conceivable," he says, "that someone might possibly have escaped from their singing; but from their silence certainly never."[1]

That more fatal weapon, silence, is, perhaps, the key to the continuing ambiguity in the interpretation of the Second Amendment.

If the founders had wanted to guarantee "the right of the people to keep and bear Arms" they could have done so in a completely transparent manner: "Congress shall make no laws that infringe on the right of the people to keep and bear arms." They did not. Why not? Why did they remain silent on so important an issue?

What they did say was: "A well regulated Militia, being necessary to the security of a free State, the right of the people to keep and bear Arms, shall not be infringed."

The most natural and unstrained interpretation of this statement is that it concerns the necessity for a "well regulated" militia (a pressing issue at the time) and forbids the government from disarming the militia.

1 Kafka, Franz (1946). *The Great Wall of China: Stories and Reflections*, New York: Schocken Books.

The argument for the right to keep arms in one's home for self-defense is persuasive; this seems to have been an accepted right, with a few exceptions, in colonial times. But this right is not embedded in the Constitution in the Second Amendment. Madison rejected a proposal that it be included. Why?

Perhaps, part of the answer is provided by the fact that many communities in colonial times restricted arms in various ways. As Justice Breyer states in his dissent to the Supreme Court decision of 2008:

> ...colonial history itself offers important examples of the kinds of gun regulation that citizens would then have thought compatible with the 'right to keep and bear arms,'... Boston, Philadelphia, and New York City, the three largest cities in America during that period, all restricted the firing of guns within city limits to at least some degree... Boston in 1746 had a law prohibiting the 'discharge' of any gun or pistol... in the Town on penalty of 40 shillings... Philadelphia prohibited on penalty of 5 shillings (or 2 days in jail if the fine were not paid) firing a gun or setting off fireworks in Philadelphia without a 'governor's special license.' And New York City banned, on the penalty of a 20-shilling fine, the firing of guns (even in houses) for three days surrounding New Year's Day... Rhode Island [prohibited] ... 'the firing of any gun or pistol in the streets of any of the Towns of this Government or in any Tavern of the same, after dark on any night whatsoever'... [2]

In colonial times, it was thus accepted that communities have the right to restrict arms to protect public safety. Creating a constitutional right to keep and bear arms would have invalidated already existing local laws.

It is just this right that Madison rejected, possibly reasoning that, in this situation, the Constitution should not guarantee a right that was already restricted by local communities. Therefore, may we conclude that the silence of the Constitution on this issue is,

2   District of Columbia *et al.*, Petitioners *v.* Dick Anthony Heller, 554 U.S. No. 07–290.

in effect, an affirmation of the fact that the right "to keep and bear Arms" as a right unrelated to the militia was deliberately not included in the Second Amendment to the Constitution. In this case, perhaps, song would have been preferable to silence.

11

Introduction to *Copenhagen*

In his play Copenhagen, *Michael Frayn created a dramatic work with references to some of the most pressing issues in twentieth-century physics. But does knowledge of physics help in an understanding of the play?*

This essay is based on program notes written for Trinity Repertory's production of the Michael Frayn play Copenhagen, *directed by Oskar Eustis in December of 2002.*

O f course Michael Frayn's *Copenhagen* is not about physics. It is about the memories and interactions of three important, passionate, figures struggling to make sense of a meeting that took place in Nazi-occupied Denmark in September of 1941. The question that dominates the play is why did Heisenberg, head of the German nuclear project, make that visit to Bohr, his old mentor, one of the creators of quantum theory and nuclear physics.

Since memory of the past is shaped by the present, and since each of the characters, particularly Heisenberg, has reasons of their own for wanting to reinterpret what their intentions were, we probably never can be sure what the actual reason for the visit was. (According to the quantum superposition principle, might it have been a combination of all of the various possibilities – all of the above, as the multiple-choice question puts it.)

But two of the three characters, Bohr and Heisenberg, happen to be among the founders of modern quantum theory – certainly one of the great intellectual achievements of the twentieth century. Further, the play makes liberal reference to various concepts of physical theory: matrix and wave mechanics, uncertainty, critical mass, uranium 235, uranium 238, etc. Does some knowledge of these aid either the actors or the audience? This question was the heart of a course (Physics 10) taught at Brown University during the spring term of 2002. The participants in the course (in addition to about 60 very gifted Brown University students) were the actors performing tonight, Anne Scurria, Stephen Thorne, and Timothy Crowe, as well as Professors Thomas Beirsteker, Abbot Gleason, Oscar Eustis, and myself.

In the class meetings the actors read through the play as they would in a "table rehearsal" directing any questions involving science, politics, or history to either professors or students. One result was that what physics was discussed was not presented in the usual logical or historical sequence – from beginning to middle and end – but rather starting from questions raised by the play and working forward, backward, and sideways to clear up "loose ends." The result was disjointed from a traditional view, but generated a level of interest rarely displayed in science classrooms: e.g., "You're not getting out of this room until you tell us whether Schrödinger's Cat is alive or dead."

The actors, students, and professors agreed that their understanding of and ability to interpret the play was substantially improved by participation in the course. But we cannot expect that you will interrupt your busy lives to repeat our experience. However, without directing you to a physics text, it is useful for you to know that uranium 235 and 238 are two uranium isotopes (different in mass, but chemically identical – therefore very difficult to separate from one another). Uranium 235, the isotope used in a bomb, comprises only about 1% of natural uranium. On the other hand, plutonium 239 (also bomb material), which can be made from uranium 238 in a reactor, can be chemically (thus much more easily) separated from uranium. Therefore making a reactor might be an attempt to produce power or a step toward a bomb. Was it then a rational decision to conclude that under wartime conditions it might have been impossible for Germany to enrich enough uranium 235 for a bomb? (The U.S.A. with our massive resources required about two years to produce enough enriched uranium 235 to make one bomb, dropped, untested, on Hiroshima.) What was Heisenberg really trying to do? Was he sure himself?

But the play's the thing. Although participation in Physics 10 helped tonight's actors to better understand their roles, does their increased knowledge deepen their interpretation and result in a profounder performance? And most important, does this make the play more meaningful? You, the audience must decide. In any case – enjoy the play.

12

The Unpaid Debt

Money is required to do fundamental scientific research, but the potential benefits of this research are often not clear in advance, even though their worth in terms of contributions to the gross domestic product, historically, have been enormous. Who should pay for fundamental research and why?

This essay is based on an article originally published in Nature Physics on December 1, 2007.

Preparing for a talk on the fiftieth anniversary of the Bardeen–Cooper–Schrieffer (BCS) theory of superconductivity I was struck by a footnote on the first page of our 1957 paper:[1] "This work was supported in part by the Office of Ordnance Research, U.S. Army" – a program officer whose mission might have included improving artillery shells found it appropriate to fund a project in fundamental science. This spurred me into reflection on funding for scientific research, then and now.

Money is required to do science and, as systems become more complex, more people, equipment, and therefore more money is required for each new result. Naturally, people hark back with sentimentality to the good old days when results could be obtained on a tabletop. In fact, some results are still obtained on tabletops, but the tables are getting larger and the tops more expensive. More

---

1 Bardeen, J., Cooper, L. N, and Schrieffer, J. R. (1957). Theory of Superconductivity, *Physical Review*, **108**, 1175–1204.

and more results come from huge collaborations demanding enormous resources. And this brings us inevitably to the questions of who pays, how does one pay, and why should one pay.

It is commonly accepted that fundamental research provides the basis for technology of the future. For superconductivity this is demonstrably so. In addition to the profound influence that the BCS theory has had on fundamental science (for instance, with its seminal introduction of spontaneous broken symmetry, an idea that is at the foundation of the Standard Model, and of pairing, which plays an important role in nuclei, neutron stars, helium 3, and dense quark matter), practical applications are close to ubiquitous: power transmission, electronics, magnetic resonance imaging (MRI), and possible quantum computing. The list continues.

Some of these applications could have been foreseen in 1957, but most could not. The technological consequences of fundamental science are largely unpredictable and often the most important are the ones we cannot foresee. This is a problem faced by government agencies when funds are tight; it is a problem faced by corporations when they have to report to shareholders. Everyone will agree that the invention of radio, radar, the transistor, penicillin – let me not list them all – have a value that is hard to overestimate for the quality of our lives as well as the gross domestic product (GDP). Further, it is generally agreed that these could not have come about without advances in fundamental science (electromagnetic theory, quantum theory, and the microbe theory of disease) that came before. No one questions the enormous value of the intellectual property given to us by fundamental science.

But we have no mechanism to protect this property and to reap its economic benefits. (It is sometimes argued that science only discovers what is already there. I have argued elsewhere that science can be regarded as invented as much as discovered.[2]) And it sometimes takes a long time to go from fundamental science to

2 Cooper, L. N (1968). *An Introduction to the Meaning and Structure of Physics,* New York: Harper & Row, pp. 71–92.

technology. Since there is no mechanism for fundamental research to pay for itself, who is to pay? One has to decide as a stockholder or a taxpayer whether or not to support research; one has to decide whether or not to spend one's own money for a program whose consequences and whose benefits are shared by many, including competitors, and are hard to predict.

I recall making the point, at a symposium organized by the Army Research Office, that the research on superconductivity in which I had participated had been financed by the Army and, as a major consequence, led to the development of what is called the super-conducting quantum interference device (SQUID – a device that can be used to make very sensitive magnetic-field measurements). The prime military user of the SQUID, as far as I know, is the Navy, which is very interested in the measurement of magnetic fields. I'm sure the Army doesn't begrudge its sister service this great benefit of research the Army financed, but it does make the problem clear. In a market economy, we expect to be paid for what we do and we expect to enjoy the fruits of our investments. When one invests in research – other than for intellectual pleasure, for which resources are somewhat limited – one invests statistically, based on history that tells us the benefits will be enormous. In times of fat budgets, perhaps this is not too much of a problem; but when budgets are constricted, one always tends to cut the future for the present, to meet short-term rather than long-term goals. In recent years, for example, in spite of repeated promises to the contrary, the real dollar budget for basic research has been reduced, putting great strains on the research community.

It would have been difficult to predict that the investigations of Maxwell, Lorentz, and Einstein in electromagnetic theory would lead to improvements in communications. Or that Kamerlingh Onnes' work on superconductivity would one day help us build better electronics. Few would have expected that Schrödinger and Heisenberg's quantum mechanics would lead to the transistor and computers, that Townes' work on millimeter radiation would give us laser surgery. Or that Bloch and Purcell's solid-state research would lead to MRI. Premature targeted programs to obtain these technologies would have failed. Worse, resources would have been taken away from the scientists who in fact made them possible.

In 1887 Edward Bellamy wrote with a certain optimism (viewing his own time from the utopian future he was visiting):

If we could have devised an arrangement for providing everybody with music in their homes, perfect in quality, unlimited in quantity, suited to every mood, and beginning and ceasing at will, we should have considered the limit of human felicity already attained, and ceased to strive for further improvements.[3]

If universities and her Majesty's Office of Royal Navy Research had been instructed, as an imperative social objective, to provide every home with music and had directed all of their research funds to obtain this result as soon as possible, they would have been unlikely to have funded Maxwell, Lorentz, Einstein, and all the others who have made music in people's homes indeed possible. We might have developments of then existing technology such as player pianos or elaborate music boxes rather than the stereo equipment found today in almost every home.

To my mind, these are some of the underlying problems involved in supporting the fundamental research that is required if science is to progress. The following measures would improve the current system.

First, invest in fundamental research as a separate line item, separated from all development projects, as some fixed percentage of the GDP, and thought of as a payment – a type of royalty – on the economic worth of fundamental ideas of the past as well as an investment for the future. This payment should be regarded in the same way as any other obligation (interest on the national debt,

3 Bellamy, Edward (1888). *Looking Backward: 2000–1887*, New York: Random House, p. 90.

Fig. 12.1 This image began as a drawing by Pink Floyd drummer Nick Mason, and was used as the cover art for their 1971 album, *Relics*. Visual artist Storm Thorgerson made a model based on the drawing and gave it to Mason as a gift. A photograph of the model was used on the CD reissue of *Relics* in 1994. Photograph used by permission of Nick Mason.

for example). It would be a transfer from scientists who have created value in the past to those who will create it in the future. Fundamental research would not be subjected to momentary political whims; it would be an obligation rather than a discretionary item on the budget. One could reasonably predict what was available for fundamental research, so as to plan in some sensible way for the future. It takes years to produce a Ph.D. In deciding whether or not to enter into a career in science one could make some estimate of the economics at the end of the path.

Second, the funds should be distributed in a manner that maximizes creativity – certainly not through one super-agency. I would hope to have the funds distributed among many different agencies such as the National Institutes of Health and the National Science Foundation. I would bring the military organizations – the Army Research Office, the Office of Naval Research, and the others – back into the business of supporting fundamental research. In addition there could be distributions of funds to private foundations that have shown wisdom in supporting research. The goal should be a maximum of diversity and a maximum of different types of risk-taking. In effect, one wants a highly diversified portfolio of support for fundamental research since no one can predict which directions will be most fruitful.

Third, there should be a clear distinction between development and fundamental research. One should also segregate very large projects with substantial political support from constituencies and regions from individual research projects that often have little political support. I would recommend that the large projects have a given percentage of the budget allocated to them; one could make decisions based on priorities decided among these projects. However, some portion should be reserved for relatively small individual projects. For these, there should be a minimum of micromanagement. We should put money on our best horses and let them run their race.

No single method can solve all of our problems, but the measures outlined above would substantially improve our present system. I would hope that they would make it easier for some current gifted program officer to reach as wise a decision as was made in the Army Ordnance Office fifty years ago.

Thought and Consciousness

Part Two

Thought and Consciousness

13

Source and Limits of Human Intellect

The brain is believed to be the physical source of thought. But can the human brain understand itself? Are there limits to human intellect? If so, what are they and how will we know when we have reached them?

This essay is based on an article originally published in the journal Daedalus, *109(2), in the spring of 1980.*

1

Not so long ago the eye was thought to be a somewhat miraculous organ functioning in a more prosaic body. We smile indulgently at the naïveté of our intellectual grandparents. Today, though we regard our eyes with great respect, few attribute magical properties to them. The same might be said for kidneys, the heart, and other organs. We appreciate their importance, we may understand how they work, we may not be able to build them as efficiently as nature does, yet we hardly regard them as mysterious.

The same calm does not seem to prevail when we consider the brain. Although the brain could be regarded in the same way we regard the heart, the eye, or a muscle, the functions associated with the physical entity "brain" such as thought, consciousness, and awareness of self – those most precious human characteristics – are not as easily attributed to the earthy material in which they may or may not originate, as the function of pumping might be attributed to the heart.

The brain has not always been thought to be the physical origin of thought. Aristotle identified heart as the seat of intellect, assigning to brain the function of a cooling system. We are fond of dismissing Aristotle with an ironic smile; in his defense, however, we might note that, whether or not in any individual case the brain is used for thought, it always cools. Almost one third of a normal adult's energy (about 600 K calories/day) is used to power the brain. (Whether this varies with thinking is not known.) The dissipation of 600 K calories/day is accomplished by an enormous capillary network that functions, in fact, as a remarkable cooling system.

Let me quickly say that I am not in the camp of those who tell us with a sneer that we are constructed of only $1.05 worth of raw materials. (The price by now is surely substantially higher.) Rather, I would agree with Shakespeare:

What a piece of work is a man! how noble in reason! how infinite in faculties! in form and movement how express and admirable! in action how like an angel! in apprehension how like a god! the beauty of the world, the paragon of animals![1]

Those who deny this show a meanness of mind that makes them ineligible for appointment to any humanities department; and, what may be more important, they deny the facts. Those who agree with the poet might conceal it to be granted tenure in the "hard" sciences. Worse, they often become immersed in a mysticism that embarrasses even their colleagues.

There is no difficulty "explaining" the marvel that is man by assuming the existence of a substance – call it marvelous – a substance sometimes denoted today as a "holon," or in the past, as a vital force, responsible for his unique behavior. These are what I call "Coca-Cola theories." Like the homunculus of Goethe – the little man inside the large one – the secret syrup locked in a safe

---

1 Shakespeare, W. (ca. 1600). *Hamlet*, Act II, sc. ii, lines 319–323.

in a far-off country, its essence known only to a few high priests (who profit from it), is the mysterious substance that gives virtue and being to the whole. And all of these hidden substances may exist. We have no hard evidence to the contrary. It may be, as is sometimes claimed, that by their very complexity, large systems – in particular biological systems – attain an entirely new principle of operation (a law of nature, so to speak) that is intrinsically associated with their largeness and complexity. It may even be that a modification of the quantum theory is needed to provide a special place for consciousness.[2]

It was not possible to construct the nuclear atom with the concepts and materials available in classical mechanics and electrodynamics. But the effort was priceless, leading finally to the quantum theory. It has been possible to construct the gene from materials provided by chemistry and physics. We cannot say in advance whether the existing body of science is sufficient to encompass any new experience we encounter. However, the history of assumptions such as vital forces, distinctions between organic and inorganic, or celestial and earthy materials should be enough to warn us that grandiose and unwarranted assumptions often precede careful analysis of the implications of what already exists.

The object of scientific endeavor, as I understand it, is to construct from the materials and concepts (laws of nature, if we insist) provided by the physicist all of the entities of the world including brain and mind. We do not know if this will be possible. But evidence that it is not possible would be among the more important discoveries of science. If it is possible, all of the marvel that is a human being, the distinction between myself and other, arises not from a new material or a new principle, but rather by an extraordinary organization of ordinary materials according to fundamental principles applicable in general to living and nonliving situations.

2 In Cooper, L. N (1976). How Possible Becomes Actual in the Quantum Theory, *Proceedings of the American Philosophical Society*, **120**(1), 37–45, I have argued that this is not necessary.

But what are the consequences and the implications for us if such explanation turns out to be possible? It is not an entirely comfortable prospect. The possibility that we can be explained, that we are constructed of dross and earthy material, that we contain no vaporous celestial substance, certainly makes us uneasy. Perhaps it is this that troubles Victor Weisskopf. Does removing the mystery destroy something precious? Does explanation destroy enjoyment? What is there "about experiencing these phenomena that lies outside science"?[3]

We must, of course, distinguish science as observation and explanation from experience itself. Knowledge and understanding, even understanding of experience in terms of physical phenomena, is not a substitute for the experience itself (as Goethe tells us in the first few lines of *Faust*), for experience involves nerve endings as well as the brain, whereas understanding engages the brain alone. This, it seems to me, is certainly not a limitation on science, nor should it be shocking or disappointing. We can enjoy wine and food even in the possession of a complete knowledge of the chemistry of taste and digestion. Notre Dame is distinguished not by the quality of its stone and mortar but by the remarkable organization of these materials.

In this essay I will argue that humans can be thought of as a perhaps remarkable organization of earthy material and that a "scientific" understanding of what we are and how we function is possible. And it is just the limitation on human intellect and imagination implied by such a scientific understanding of precisely how we function that I should like to explore.

2

Probing the source and nature of mind presents us with more than conceptual trauma. The brain is a complex and baffling instrument

3 Weisskopf, V. F. (1975). The Frontiers and Limits of Science, *Bulletin of the American Academy of Arts and Sciences*, 28(6), 15–26.

that has so far guarded its secrets well. It is composed of vast numbers of cells called neurons (10^{11} is an estimate commonly given for humans) held together, fed, and cleansed by various supporting structures, blood vessels, and glial ("glue") cells. It is thought that the information-processing and storage functions of the brain are accomplished by the neurons and that the other tissue is occupied primarily with housekeeping.

Speaking pictorially, neurons look somewhat like trees, and a network of neurons looks somewhat like layers of trees one on top of the other. Information in the form of electrical voltages and currents is transmitted from the axons (branches) of one layer of neurons to the dendrites (roots) of the next layer. The electrical information in the dendrites of a particular neuron is gathered together to produce a voltage across the cell body (the base of the tree), and this voltage determines the electrical signal sent along this neuron's axon (the trunk of the tree) to the ends of all of the axon branches. The ends of the axon branches are very close to the dendrites of the next layer of neurons; and the information contained in the electrical signal that reaches the tips of the axon branches of one neuron is transmitted chemically and sometimes electrically across the small spaces (the synaptic clefts) between the ends of the axon branches and the dendrite roots of the next neuron. This produces electrical voltages and currents in the dendrite roots, and the process is repeated. Thus the information flow continues.

Although the properties of individual neurons are relatively well understood, the manner in which large interacting networks of these nerve cells produce mental activity remains almost a complete mystery. This is due in part to the complexity of the central nervous systems of higher animals and to the great difficulty of observing these systems without destroying them. It also seems likely that higher central nervous system properties are of unusual subtlety involving small changes in the activities of large numbers of neurons.

The means, for example, by which large numbers of neurons can interact to perform such functions as storage and retrieval of information remain unknown. Many ways to store and retrieve information exist: filing cabinets, libraries, and computers. But the fact that an animal's memory is held in a living structure and is successfully utilized, even though the animal may have no idea where his memories are stored or how they are ordered, places special requirements on theory. Current computer memories, for example, are made of elements in which yes/no information is recorded and that can be recalled by addressing the location of an element. These computers perform sequences of elementary operations with incredible speed and accuracy, completely beyond the capability of living cells. A basic problem in understanding the organization of memory in a biological system is to understand how a vast quantity of information can be stored and recalled by a system composed of vulnerable and relatively unreliable elements, with no knowledge of how or where the information has been filed.

In recent years it has come to be believed that the storage of memory in an animal central nervous system is distributed rather than local (more like a hologram than a photograph), and that such distributed memory is stored over large regions of the neural network by small but coherent modifications of large numbers of synaptic junctions.[4] Neural networks have been constructed that

4 Anderson, J. A. (1970). Two Models for Memory Organization Using Interacting Traces, *Mathematical Biosciences*, **8**, 137–160; Anderson, J. A., Cooper, L. N, Freiberger, V. V., Grenander, U., and Nass, M. (1972). Some Properties of a Neural Model for Memory, *American Association for the Advancement of Science Symposium*, December 1, 1972; Cooper, L. N (1974). A Possible Organization of Animal Memory and Learning, *Proceedings of the Nobel Symposium on Collective Properties of Physical Systems*, B. Lundquist and S. Lundquist (eds.), New York: Academic Press; Kohonen, T. (1972). Correlation Matrix Memories, *Institute of Electrical and Electronic Engineers Transactions on Computers*, **C-21**, 353–359; Longuet-Higgins, H. C. (1965). Holographic Model of Temporal Recall, *Nature*, **217**(104); Pribram, K., Nuwcr, M., and Baron, R. (1974). The Holographic Hypothesis of Memory Structure in Brain Function and Perception, in *Contemporary Developments in Mathematical Psychology*, vol. 2, D. H. Krantz, R. C. Atkinson, R. D. Luce, and P. Suppes (eds.), San Francisco: W. H. Freeman, pp. 416–457; and Willshaw, D. J., Buneman, O. P., and Longuet-Higgins, H. C. (1969). Non-Holographic Associative Memory, *Nature*, **222**, 960–982.

can organize themselves in such a way that they acquire and store distributed memories. They display, on a primitive level, features such as recognition, association, and generalization, and suggest some of the mental behavior associated with animal memory and learning.

The mechanisms employed seem to be plausible biologically and are not inconsistent with known neurophysiology. In addition, the networks that result seem to be a reasonable outcome of evolutionary development under the pressure of survival.[5] Although the concept of distributed transformations and memory storage is less familiar than that of local storage, distributed transformations have been discussed and analyzed, and some such transformations have probably already been observed in the central nervous system of a higher animal.[6]

It is now commonly believed that much of the learning, and resulting organization of the central nervous system, occurs owing to some kind of modification of the efficacy or strength of at least some of the synaptic junctions between neurons, thus altering the relation between presynaptic and postsynaptic potentials. It is known that small but coherent modifications of large numbers of synaptic junctions can result in distributed memories such as those mentioned above.

5 A more detailed discussion of some of these ideas is given in two articles: Anderson, J. A. and Cooper, L. N (1978). Biological Organization of Memory, in *Pluriscience,* Paris: Encyclopaedia Universalis, pp. 168–175; Cooper, L. N. (1979). Distributed Memory Storage in the Central Nervous System: Possible Test of Assumptions in Visual Cortex, *Cerebral Cortex Colloquium*, Woods Hole, Massachusetts, April 29–May 4, 1979.

6 An example is a part of the brain related to the visual system known as superior colliculus; see Mcllwain, J.T. (1976). Large Receptive Fields and Spatial Transformations in the Visual System, in *International Review of Physiology: Neurophysiology II*, vol. 10, R. Porter (ed.), Baltimore: University Park Press. Although the neurons that lead to the colliculus form a very precise, fine-grained map, cells later in the system, physically a few millimeters below the very precise cells, respond to stimuli over a wide area of visual space. Thus we have the apparently paradoxical situation – which seems to be true of other parts of the brain as well – that great precision of response is generated by systems composed of cells that progressively show less and less selectivity as the motor output of the system is approached.

Although this is not the place to go into technical details of the neural networks that have been constructed, I should like to suggest some of the basic ideas. In a distributed memory it is the simultaneous or near-simultaneous activities of many different neurons (the result of external or internal stimuli) that are of interest. Thus a large spatially distributed pattern of neuron discharges, each of which might not be very far from spontaneous activity, could contain important, if hard to detect, information.

Each neuron communicates electrically with one or many others. To illustrate some of the important features of such systems, consider the behavior of an idealized neural network that might be regarded as a model component of a nervous system. Consider a large number of neurons that are connected to some other large number of neurons, so that on the average all neurons are connected to many others.

The actual synaptic connections between one neuron and another are generally complex and redundant; let us idealize the network by replacing this multiplicity of synapses between axons and dendrites by a single ideal junction that summarizes logically the effect of all of the synaptic contacts between an incoming neuron to an outgoing neuron. For utter simplicity, imagine that each of the incoming neurons is connected to each of the outgoing neurons by a single ideal junction.

We may then regard the synaptic strengths of the ideal junctions as a transformation that takes us from the incoming neuron activities to the outgoing activities. This transforms the neural activities incoming into the neural activities outgoing. It is in such modifiable sets of synaptic junctions that the experience and memory of the system are proposed to be stored. In contrast with machine memory that is, at present, local (an event stored in a specific place and addressable by locality, requiring some equivalent of indices and files), animal memory is likely to be distributed and addressable by content or by association. In addition, for animals there need be no clear separation between memory and "logic."

Each synaptic junction stores some portion of the entire experience of the system, as reflected in the firing rates of the neurons connected to this junction. Each experience or association, however, is stored over the entire array of junctions. This is the essential meaning of a distributed memory: each event is stored over a large portion of the system, while at any particular local point many events are superimposed.

The fundamental problem posed by such a distributed memory is the address and accuracy of recall of the stored patterns. It has been shown that such non-local mappings can serve in a highly precise fashion as a memory that is content-addressable and in which "logic" is a result of association and an outcome of the nature of the memory itself.

In some circumstances a distributed memory might be "confused," in the sense that it will respond to new events as if they were seen before, if the new event is similar to ones that have been previously seen. It will "recognize" and "associate" events never, in fact, seen or associated before.

The memory will tend to categorize stimuli on the basis of the past history of the system. If the system has learned to identify many items sharing some common feature, some new item possessing this feature can eventually be recognized more strongly than any particular event actually presented. This, of course, is reminiscent of psychological properties called "generalization" or "abstraction." From such a point of view, generalization grows from the loss of detail of individual instances, a characteristic of distributed systems. Something like this seems to occur in a few psychological contexts where it can be checked.

We have here an explicit realization of what might loosely be called "animal logic" – which, of course, is not logic at all. Rather, what occurs might be described as the result of a built-in directive to "jump to conclusions." The associative memory by its nature goes from recognition of particular items, each possessing some common characteristic, to the recognition of the characteristic itself, which can be described in language, for example, as passing from particulars: cat^1, cat^2, cat^3 – to the general: cat. How fast this step is taken depends on details of the system.

In addition to "errors" of recognition, the associative memory also makes errors of association. If, for example, all (or many) of the items of some class associate some particular item, a new event that is also in the class will not only be recognized, but will also associate as strongly as any of the items explicitly contained in the class.

If errors of recognition lead to the process described as going from particulars to the general, errors of association might be described as going from particulars to a universal: cat^1 meows, cat^2 meows, cat^3 meows → all cats meow.

There is, of course, no "justification" for this process. Whatever efficacy it has will depend on the order of the world in which the animal system finds itself. If the world is properly ordered, an animal system that "jumps to conclusions" in the sense above may be better able to adapt and react to the hazards of its environment, and thus survive. The animal–philosopher sophisticated enough to argue, "The tiger ate my friend but that does not allow me to conclude that he might want to eat me," might be a recent development whose survival depends on other less sophisticated animals who do jump to conclusions.

By a sequence of such transformations, a fabric of events and connections is woven that is rich as well as suggestive. We can see the possibility of a flow of electrical activity influenced both by internal distributed transformations and the external input. This flow is governed not only by direct associations (which can be explicitly learned), but also by indirect associations due to the overlapping of the mapped events. We can easily imagine situations arising in which direct access to an event, or a class of events, has been lost while the existence of this event or class of events influences the flow of electrical activity.

In constructing the means by which such mappings can be attained by a neural network in interaction with an environment, it has been necessary to utilize a distinction between forming an internal representation of events in the external world and producing a response to these events that is matched against what is expected or desired in the external world.

We use the idea that the internal electrical activity that in one mind signals the presence of an external event is not necessarily (or likely to be) the same electrical activity that signals the presence of the same event for another mind. There is nothing that requires that the same external event be transformed into the same neural patterns by different animals. What is required for eventual agreement between animals in their description of the external world is that the relation of the signals to each other and to events in the external world be the same.

Omitting technical details, what seems most important is that such neural networks can generate their own response to incoming patterns in such a way as to construct distributed transformations that can function as memories capable of recognition and association, and that to a limited extent, these transformations can be regarded as internal representations of what has arrived from the outside world. The "logic" of the flow of electrical activity in such systems is a consequence of the interaction of the neural network, obeying rather general directives, with the particularity of the environment. These systems are self-assembling and require a minimum of pre-built higher-level mental structures.

We are some distance from an explicit theory for the physical basis of mental activity. But, surely, our thinking originates in the material in our head. Almost certainly, what we call thoughts or ideas are the play of electrical activity in neural networks. Very probably these networks modify themselves with experience to store memory. And, quite possibly, as has been implied, an underlying directive is to seek patterns, associations, connections in the external world – to jump to conclusions.

Obviously such a procedure is fraught with difficulties. Mistakes of observation and analysis will be made; but it is a procedure with survival value in a world that is sufficiently ordered, for in an ordered world it is advantageous for an animal to learn quickly and to act on what has been learned.[7] This animal can alter its behavior in days or minutes (rather than waiting generations for genetic changes) to adapt to different circumstances. Mistakes will be made, and these will sometimes be dangerous. But to the degree that the world is truly ordered, it is to our advantage to learn fast. If this is the primitive directive nature has given to brain, what are the implications for the knowledge it can achieve?

3

Surely one of man's oldest questions concerns the nature of knowledge. What is it? How do we acquire it? How can knowledge be distinguished from opinion or belief? How can "truth" be distinguished from "untruth"? These questions were asked by Socrates and Plato, and though we ask again, I do not believe that we will answer them here.

From the point of view of a strict empiricism, the attempt to go beyond sense data (without the introduction of assumptions or concepts not justified by the data themselves) seems to fail. Though it was surely a shock for the philosophical world to discover that the knowledge we want cannot be obtained purely empirically, this is not necessarily because nothing beyond sense data exists: rather, it seems reasonable to conclude that such a limitation on permissible assumptions is too rigid, that with such a limitation no construction is possible. It seems to me that this is a single example of the more general problem involving all scientific knowledge. Assumptions must be made (as Laplace, we make no more than we

---

7 Whether the world is truly ordered and whether this order will continue is related to the problem of induction. Can we go from N to $N + 1$? We do not have to answer this question. For our argument it is sufficient that our own experience was ordered in the past. We can show that it is highly improbable that no order existed in the past for at least some portion of our own experience.

need), and the resulting structures judged by their coherence, consistency, elegance, and by their agreement, eventually, with sense data. The assumptions we do make – even the most primitive, those concerning space, time, and causality – need not be thought of as innately given or as due to some categorical imperative, but rather as assumptions introduced by a relatively unbiased mind to produce the most efficient ordering of the phenomena in the particular world in which we live.

The proposal that knowledge arises from a neural network interacting with an ordered environment can then be regarded as an attempt to construct a theory of the nature and limitations of human knowledge in a manner similar to the construction of theories in other domains. Such a theory with its assumptions and consequences forms a structure that is or is not in correspondence with our experience of knowledge itself. We therefore come to know the source and nature of knowledge in precisely the same way we come to know other phenomena. We judge theories of knowledge just as we judge other theories: by their consistency, explanatory power, and agreement with all manners of experience.

Since in such an approach we introduce entities that may or may not be directly observable, and these entities supposedly exist within the head, we can be accused of a sort of mentalism. But this, not by a misplaced positivism, is precisely how science progresses. Science has always constructed little machines or mechanisms "behind immediate experience" that connect various phenomena. The true "mechanism" of science is not a physical but a logical machine whose wheels and gears can be manipulated and which responds to pushes and pulls on the various levers in a consistent manner and in agreement with experience.

In an ordered environment the neural networks I have mentioned do to a certain extent form an internal representation of the external world (as has been demonstrated analytically and by computer simulation). Events that are common and associated in the external world are recognized (produce characteristic electrical activity) and associated in correspondence with what occurs in the world. These recognitions and associations are, it seems to me, the biological primitive of what we eventually call knowledge. Of course, the "knowledge" so gained is highly suspect; much of the time the generalization leapt to will be misleading or incorrect.

The painful and long schooling we are subjected to may be regarded as an attempt to discipline the built-in and somewhat freewheeling biological directive. Bertrand Russell once pointed out that we are born with belief. We must be trained to be skeptical. We see all about us and repeated over and over generalizations that are too quickly made, attempts to explain too much with too little. (These paragraphs may be an example.)

Conspiracy theories abound. It is sometimes more difficult for us to accept a single chance event than to concoct some incredible conspiracy to explain it. The discipline we impose on ourselves is directed, in part, to separate these somewhat fantastic concoctions from what we regard as legitimate theory. But there is no clean way to do it. Conspiracy theories can never be completely defeated, only made less plausible. And legitimate theories can never be shown to be "true" (nor can they be "falsified" any more than one can falsify conspiracy theories). They can only be made consistent, elegant, economical, and in agreement with observation.

If the source of human intellect is in such a piece of biological machinery – machinery directed by nature to act to aid survival, machinery whose use has been refined and perverted (as we have refined and perverted eating and sex), mental machinery that has often made body a slave to its demands – what, with this limited imperfect apparatus, so bounded in size, duration, accuracy, and speed, can we achieve?

There are obvious physical limits of intellect owing to the biology and chemistry of our mental apparatus. But these are rapidly being extended. The age of machine-assisted intellect has already dawned, and its outcome can hardly be guessed. Human limitations of memory, speed, accuracy, of computation, and of ability to catalog, recall, and compare are being, and will continue to be, dramatically extended. The electronic computer is for the mind what the bulldozer and steam hammer, in extending its range, power, and speed, are for the arm.

It therefore seems to me that limitations imposed on us by physical constraints on the machinery given to us by nature (while they may be significant when such ultimate physical limitations as the speed of light are encountered – somewhat remote at present) are not really the limitations that are of interest, since they can be extended by the use of machines to assist us.[8] Assuming that we are allowed essentially unlimited (machine-assisted) memory, speed of recall, and speed of sequential calculation, what are the limits of our intellect? Are there any situations for which order exists that we cannot penetrate to discover or comprehend that order? Is there any limit to our capacity for understanding?

A theory, an explanation, understanding comes about (at least as we visualize it as present) as a structure of relations connecting events; assumptions with consequences, forces with motion, motion with observed position, or forces with atomic orbits, with atomic transitions, with emitted (and observed) light, and so on. The full structure of even the accepted and classical theories is too complex to fit into any one mind at any given time. Although physicists would unanimously agree that the uranium atom with its ninety-two electrons is understood in principle, no one has come close to calculating completely all of the details. In such a situation the underlying rules (in this case the quantum theory and electrical forces), a few very well worked-out examples in complete agreement with experiment (the hydrogen atom), and a sense that the complex situations are understood (at least qualitatively – no surprises, nothing that does not seem to follow if we are

---

8 It is sometimes said that computers, since they process information, are different in nature from bulldozers that move earth. It seems to me, however, that it is more appropriate to regard computers (at least as now constructed) as extensions of our own abilities, since they perform no single operation that we are not capable of doing ourselves.

willing to work in a straightforward manner) seems to satisfy us. And, indeed, if some aspect of the behavior of the uranium atom becomes of interest, a variety of techniques will be (in fact, have been) developed to obtain approximate solutions of the fundamental equations (consequences of the "fundamental rules").

We are much less comfortable if the fundamental rules are believed understood, and yet some qualitatively different phenomena cannot be seen to follow from the original rules. Such situations have occurred in physics: superfluidity and superconductivity, phase transitions and turbulence come immediately to mind. The first two turned out to be situations in which large systems displayed qualitatively new properties that were not easy to guess from the fundamental rules. In the case of turbulence, we seem to be entering into a new and intriguing area in which equations yield solutions that are more easily characterized as random (even though strictly speaking they are deterministic). This may be the case for weather prediction.

The great, outstanding biologically related problems concerning the human mind, consciousness, and self-awareness will, I believe, fall into a similar category – new, perhaps surprising qualitative properties of large and complex systems that follow rather easily (once we understand them) as consequences of fundamental biology and chemistry that is already in place.

But we might ask whether situations could arise in which even the rudimentary qualitative connections of the explanations are so complex that they cannot be seen or visualized by a human being; where, for example, an "explanation" exists only as a complex computer program, a program in which theorems are connected to others, not by a few simple rules, but by so many rules and conditions that we literally cannot see the interconnections between things beyond the fact that certain inputs produce certain outputs.

There can be no guarantee that explanations of the type we have come to expect will always be possible. Knowledge might exist that is essentially history, even more diverse and confusing than human history, with no rules at all. It might turn out that in some situations our "understanding" will be nothing more than a computer program. It may be that some phenomena are too complex ever to be understood in classical terms (perhaps economic behavior, social behavior, or the stock market).

But the scientific enterprise is so young, so successful, and has shown itself to be so adaptable to different circumstances, that it seems somewhat premature to throw up one's hands and conclude that it will falter over what probably will turn out to be problems just a little more complex – involving just a few more elements to be handled at once – than the ones already solved. Had we done so with any of the recently solved problems in physics that seemed so intractable just a generation ago, some of the most beautiful results of the century might have eluded us.

A question of this type is just the one we are considering. Can a brain understand a brain? The answer to this seemingly paradoxical question is relatively simple. If we require 10^{11} rules or if we must follow the electrical activity in 10^{11} neurons in detail to understand a brain, we are unlikely to ever get such understanding into our present head. But if, as seems much more likely, the brain, as other biological systems, is organized by repetition and elaboration of a few fundamental principles, understanding not only can be had, but also can be had in what we will regard as classical terms.

If there is a limit in this sense, it seems to me to be more one of interest than capacity. The scientific quest for understanding is often understanding of general relationships; when the situation is of more interest due to its complete uniqueness or individuality, "scientific" understanding does not become less possible but rather less interesting. It is perhaps this that most distinguishes science from the arts. An individual work of art is of interest for its particularity. For a work of art, how a thing is said is often more important than what is said; in science this is seldom, perhaps never, the case. There is very likely nothing about the experience of the Beethoven sonata that cannot be understood by science; but there is probably minimal interest in this understanding, since what is of interest is the individuality, the particularity of the experience rather than the generality of the understanding.

Putting aside essentially physical limitations, it seems to me that our intellect and understanding are bounded only by the limits of the structures we can invent, and that any limitation of intellect is a limitation of inventiveness or imagination. Is there a limit to human imagination or inventiveness? Is there a limit (other than physical) to the metaphorical, mathematical, scientific structures we can create?

It is an old and, I believe, tired conjecture that certain ideas or concepts are built into our mental apparatus. It has been proposed at one time or another as far back as Plato's *Meno* that our notions of space, time, and causality, or even certain aspects of grammar, are determined genetically. If such structures are built in genetically, they would of course be easier to learn, or perhaps to recall, but they would then be intrinsic limitations on intellect, since, being pre-built, they could not be discarded or changed very easily. (Presumably some ideas or concepts could not be learned and thus would not be accessible to human intellect.)

This idea of built-in or innate mental concepts is old and occasionally reappears in high fashion. There is no way we can be sure, at present, whether or not such higher-level built-in structures exist. The notion is difficult to define precisely, since it is evident that our learning begins from some genetic initial state. The important question is precisely this: How structured is the genetic initial state? Does it include what might be called higher-level concepts, or are the so-called built-in concepts so primitive that they amount to little more than the possible arrangements between neurons or the possible strengths of synaptic junctions.[9] The strongest

9 This is just a step from the very old question: selection versus instruction. Is what we have in our heads selected from a very large number of pre-existing possibilities, or is it put together in interaction with the environment by instruction? The answer is obviously: both. The question becomes interesting only when the level of complexity of

argument against such higher-level built-in concepts seems to me to be that they are premature and ill-defined.[10] Where they can be defined, concepts such as time, space, or geometry seem not to be built in but more easily explained as a consequence of the interaction of a relatively unbiased mental apparatus with some of the most basic ordering of our environment.

I suspect that, beyond such built-in primitives of the general nature of sucking, grasping, babbling, and a mental directive to "leap to conclusions," much of the rest is learned in interaction with the environment or explicitly taught. What evidence exists indicates that the general recognition and association capabilities of the neural network function with a minimum of bias and are as content to make one association as any other. The biological machinery has evolved in such a way as, so to speak, to be ready for anything. The apparatus that preserved us from predators on the grassy plains serves us as musicians and mathematicians. How we think is tempered and disciplined by training, not by nature.

A child seems willing to accept any connection, marvelous or trivial, if it is repeated and part of his world. No child of the television era questions the connection between the push of a button and the appearance of talking images on a screen. It would have

the elementary items from which the selection is to be made is specified. Let me give an example. Is a novel written by selection or instruction? Usually we would say by instruction. But the selectionist could argue that each letter, including punctuation and blanks, that forms the sequence that makes the entire story has been selected from, let us say, a hundred possibilities, so that a novel consisting of a million letters has been selected from the 100^{106} possibilities.

10 The problem is trivialized when one introduces assumptions of higher-level built-in mental structures before one has fully explored the consequences of far simpler, more general structures that we already know exist. Perhaps we might profit from some recent history. In theoretical physics in the last few generations many radical new assumptions have been introduced in attempts to resolve very difficult fundamental questions. Many of these have turned out to be unnecessary. Much of the remarkable progress that has been made in the last forty years has been the result of very careful and often brilliant working out of the rich consequences of assumptions already in place.

been for science fiction to propose such a thing in the nineteenth century.[11]

Whatever limits there are to human imagination would seem to be those we have placed on ourselves by culture, training, and experience. We live by the learned images and the myths in our heads; our imagination as well as our behavior is restricted by what has been learned. This seems to me to be the case even in those situations that seem superficially to be beyond human visualization: a fourth space dimension or Einstein's time. Mathematicians or physicists, of course, have no difficulty at all working with Einstein's time or any number of space dimensions. The problem for the layman (or even for mathematicians or physicists when they try to think in ordinary terms) is that visualization is taken to mean visualization in terms of concepts we have already learned, that we already have in our heads. Sometimes this is not possible. The new concept is independent of (or even contradictory to) the old.

Thus Einstein clocks and ordinary clocks cannot exist together (and both be given the name "clocks"). But we are not satisfied that something is time unless it is expressed in terms of ordinary clocks. If we are to visualize four-space dimensions, we insist that it be done using the limitations in visualizing three.

It also seems rather unlikely that anything is built-in to give us our notions of space (three-dimensional) or time. Rather, I would say that these deep and seemingly unbudgeable concepts are the result of the interaction of our relatively unbiased mental apparatus with some of the most elementary, repeated, and primitive orderings of the world we live in. I would conjecture that the same apparatus in a world of two or four spatial dimensions would quickly acquire the appropriate primitive notions. This comes close to occurring in actual experience. Engineers and physicists who design high-energy machines have developed a fingertip intuitive notion of Einstein's time and space that is applied as a matter of

11 Nor is any child of the digital age surprised by the images or responses on his iPad or iPhone.

everyday routine in their work. Yet they switch without any apparent mental anguish to ordinary concepts of space and time before they reach home.

It pleases my natural optimism to be able to conclude that in this most important respect – imagination – there is no limit to human intellect. Our imagination is marvelously free, capable of any juxtaposition, unbounded by logic or even by experience.[12] Those physical limits to our intellect such as small and inaccurate memory, slow speed of computation, and so on can and will be increased almost without bound (or at least until the limits due to physical laws such as the speed of light intervene), so that one day we may regard them as irrelevant as we now regard physical limitations of our strength. The conceptual limits we feel so painfully are not those of nature but (just as for social behavior) those of training. They are there not because of how our heads are made but by what circumstances and education have put in them. Understanding and use of new concepts or ideas requires retraining the old generation, or (better) different training for the new. What is required is a newborn head, not a newly designed head.

It also seems to me that the brain we possess is very likely capable of achieving understanding of its own construction and function; that this poses no threat to our humanness if we remind ourselves that such understanding is no substitute for experience itself. It is the interaction of brain and body with the outside world that shapes them both; what brain understands does not replace the sensations it receives. There is surely something about experiencing a Beethoven sonata that lies outside science. But there is nothing about the experience that cannot be understood by science.

Although we no doubt will find ourselves to be made of ordinary materials, we might still marvel at their incredible

---

12 Fantasy might be regarded as just such juxtapositions, a mélange of various actual or created elements of our memory, fueled by desire and, at its freest, liberated from the constraints of logic or reality. Nothing is more shattering for fantasy than the mental intrusion of such constraints.

organization. Although the entire universe might be determined, there is room for our creativity and for our individuality. What we call "I," the precious individual "I," is a result of genetic direction of the materials of which our bodies are constructed, in particular the few billion neurons protected by our skull, interacting via the sensory and motor apparatus with all of the vagaries of the environment in which we find ourselves, a precious particular consequence of universal or general "laws." The result is a completely unique object. There is no wiring diagram anywhere in the universe for what is contained in any individual head. Each head, each person, is completely individual in a sense as profound as anyone has imagined. There is no replacement for the individual experience that shapes every person. When an individual dies, he or she is lost irretrievably.

I am reminded of a distinguished physicist late in his career who was asked if he would like to be cloned. He thought he would rather not, since much of his highly successful career he felt was due to a series of happy accidents unlikely to be reproduced. "Suppose you were cloned and all the accidents were reproduced as well?" He declined again – "That would be boring."

14

Neural Networks

What are neural networks and are they good for anything? If so, what? Can they lead to machines that think the way we do?

This essay is based on an article originally published in OMNI Magazine, *11(6) in 1989.*

Once upon a time, not so long ago, neural networks were as easy to sell as Rhode Island wine in France. Reading the newspapers and trade journals, one might conclude that things have changed. To a certain extent this is the case. Scientific fashion has undergone one of its seasonal flip-flops. From the wintery opinion that neural networks could do nothing, we have, among the current summertime excesses, the suggestion that they will do everything. But the jury is still out. Commercial enterprises, sensing the possibilities and experiencing success, are beginning to move into the field. A recent study from the Defense Advanced Research Projects Agency, hailing this new technology, endorsing its potential importance and supporting applications to real-world problems, cautiously encourages benchmark tests and comparisons between neural networks and more conventional techniques.

What's all the fuss about? Put somewhat simplistically, we're talking about machines that can think. Now that kind of statement generates anticipation as well as a certain anxiety. So, to paraphrase Max Beerbohm, let's avoid excitement and approach the subject in a gentle, timid, roundabout way!

From stone axes to nuclear devices, humans have been inventing and constructing machines that enhance our ability to do better and/or more economically the productive and destructive things we do ourselves. And from earliest times men have, no doubt, worried to some extent about the consequences of these inventions. The nineteenth-century Luddites, as many an assembly-line worker since, saw new machines as eliminating their jobs. Perhaps Achilles and his fellow Achaean warriors thought in a troubled way about the killing power of newly available technology obsoleting their heroic skills. Much of this worry, we realize, is misplaced. Many of the jobs made redundant are jobs (like reading zip codes) human beings should not have been doing in the first place. Further, often, new machines create new jobs.

However, we seem to have reached a new plateau. What is proposed is that the brain itself, the source of thought, the seat of memory, the physical basis of mind, consciousness, and self-awareness, all that make us distinct and human, might be mimicked by a machine that thinks as well as we do or even better. The problem has perhaps been exacerbated by statements about computers as models of the brain. Although computers add, subtract, and execute instructions with incredible speed and accuracy, a computer is almost as bad a model for the brain as the brain is for a computer. Neural networks are much more reasonable models for components of the brain; in this lies their promise and what so many perceive as their threat.

Neural networks are inspired by biological systems where large numbers of nerve cells that individually function rather slowly and imperfectly, collectively perform tasks that even the largest computers have not been able to match. They are made of many relatively simple processors connected to one another by variable memory elements whose weights are adjusted by experience. They differ from the now standard Von Neumann computer in that they characteristically process information in a manner that is highly parallel rather than serial, and that they learn (memory element weights and thresholds are adjusted by experience) so that to a certain extent they program themselves. They differ from the usual artificial intelligence systems in that (since neural networks learn) the solution of real-world problems requires much less of the expensive and elaborate programming and knowledge engineering needed for such artificial intelligence products as rule-based expert systems. Thus, neural network systems, working with current computers, seem to some of us to represent the next generation of computer architecture: systems that combine the enormous processing power of Von Neumann computers with the ability to make sensible decisions and to learn by ordinary experience – as we do ourselves.

When interest in neural networks revived some fifteen years ago, few people believed that such systems would ever be of any use. Computers worked too well; it was felt that they could be programmed to perform any desired task. Now, the limitations of current programming methods in solving many problems involving difficult to define rules or complex pattern recognition are widely recognized, and it is believed that in these areas neural networks can contribute in a major way. The problem has become to actually incorporate such networks into systems that solve real-world problems economically.

Neural networks, as presently constructed, are made of layers of relatively simple non-linear elements. What is most interesting is the possibility that by a learning procedure, a non-linear mapping of great complexity that serves a useful function can be constructed. Although it may sometimes be that the final constructed mapping is equivalent to some known solution, what is important is the speed with which this solution can be obtained by the learning procedure as well as the ease of obtaining the solution without enlisting knowledge engineers or the complex programming of many rules.

In their current state, neural networks are probably best at problems related to pattern recognition. Some existing neural network systems can efficiently and rapidly learn to separate enormously complex decision spaces. The problem of coordinating many neural networks, each a specialist in dividing some portion of the decision space, has also been solved. It is in these areas, therefore, that the first commercial uses will appear. Products that recognize characters, assembly-line parts, or signatures, that make complex decisions mimicking or improving on human experts (such as underwriters), that can diagnose engine or assembly-line problems are in the prototype stage and/or are already fielded. One expects, further, that the pattern-recognition ability coupled with and feeding back and forth to rule-based systems (as has already been done in some simple applications) will finally result in machines that share our ability to learn and duplicate our processes of reasoning – machines that might be said to think.

The question is not whether but when.[1]

Predicting the future, as we all know, is risky. Predicting the evolution of new technology is downright hazardous. Who in the 1930s would have said that among the consequences of the uncertainty principle would be transistors, silicon chips, and the vast array of solid-state devices on which all modern computers depend? Or that superconductors would lead to extraordinarily sensitive detectors of magnetic fields now carried on many naval ships. Or in the late nineteenth century, that among the consequences of the research of Maxwell, Lorentz, and Einstein would be all that we call modern communication: radio, radar, etc.?

---

1 Neural networks and other automated pattern recognition systems are presently employed to aid in such tasks as identifying relevant events at the CERN Large Hadron Collider, and played an important part in the discovery of the Higgs boson. Learning systems have been incorporated in a variety of current systems. Perhaps the most famous is IBM's Watson, which won a Jeopardy contest pitted against several human experts. Watson is now proposed to process huge amounts of data for purposes such as medical diagnosis. Whether or not neural networks are included as part of the learning system employed by Watson is perhaps irrelevant.

One of the initial reasons for neural networks was that they did parallel processing that was thought at the time to speed up the computation. However, the speed of processors has increased so rapidly that it probably is more efficient to simulate a neural network learning system using serial processors than to build them as parallel systems. This could conceivably change if quantum computing ever becomes a reality.

Accepting this risk, I would predict that neural networks will become standard components of what we today call computers. This will likely occur in a somewhat evolutionary manner: they will encroach gradually – board by board, intelligent components, that can be trained by humans in a language humans understand, into dumb machines – somewhat like the neocortex came to dominate the reptilian brain and just as the twentieth century was the century of automobiles, airplanes, telephones, and computers, the twenty-first will be the century of intelligent machines. We will not only learn to live with them but, indeed, will wonder, one day, how we ever lived without them.[2]

2 The process predicted in the last paragraph does not seem to be the most likely at present, but intelligent machines continue to evolve.

15

Thought and Mental Experience: The Turing Test

Is it possible to construct a test that can tell us when a man made entity has achieved consciousness? What can the Turing test tell us about machines that think? What does it leave out?

This essay is based on an article originally published in the book How We Learn; How We Remember: Toward an Understanding of Brain and Neural Systems, *published by World Scientific Publishing Co., in 1995.*

Over sixty years ago, Alan Turing proposed to test the hypothesis that a machine might think in the following way. Put either a machine or a person into a closed room, and communicate from outside this room in a manner that either the person or the machine can understand. (Let us say, by typing on a keyboard.) If one poses questions to whatever or whoever is in the room and if the responses do not allow one to distinguish between human and machine, then according to Turing, we can say that the machine (or the person) thinks.

Suppose, however, the entity in the room answers all questions with "I don't know," or "I don't understand what you are saying." Human being or machine? Could be a programmed computer. Could be one of our denser colleagues. Suppose the questions concern next best moves in the game of tic-tac-toe. The inhabitant of the room answers with the right move every time. A very easy program to write and not distinguishable from a human being who doesn't have to be genius class. "That's not what Turing meant," you object. "Ask every conceivable question (or some representative subset) about any situation. If the black box answers in a manner indistinguishable from an 'intelligent' human then it passes the test."

A new problem arises. Real human beings occasionally come up with somewhat quixotic answers to questions that rub them the wrong way. Suppose then, we confine ourselves to "reasonable," "rational," or perhaps "logical" or "computable" answers.

Formulated this way we do have a difficult problem but we can, in my opinion, expect that the task will be done. In a certain sense logic and reasoning are easy. Constructing the appropriate "rational" answer to a question is a process that we already understand conceptually and can presently achieve if the universe of discourse is not too large (e.g. tic-tac-toe). Presumably, for larger problems, patience and more powerful computers are what is required.

The seductive computer metaphor for brain is based on the fact that any logical process can be replicated by a well-defined sequence of simple steps, each of which can be executed in a variety of mechanical or electronic devices. Whether the software (the actual sequence of steps) or the hardware (transistors etched into wafers of silicon or living neurons inside a skull) are the same as that of the brain is regarded as irrelevant. As long as inputs and outputs match, the precise internal workings don't count – a point of view ideally suited to be challenged by Turing's test.

The problem becomes to design an actual sequence of instructions (a program) that can be executed at breathtaking speed on that object of fascination, the computer (misnamed due to its arithmetic origin – however "instruction executer" is a bit cumbersome). A delicious task for the unshaven persons in the backroom who love to hack. The program of so-called artificial intelligence (a moniker that really stuck) is then to reproduce various human activities by just such sequences of instructions so that the input–output to and from the object of worship becomes indistinguishable from (or even exceeds) that of a human being.

Thus, chess and checker playing programs, programs that paint walls, stack blocks and so on, ad-Ph.D.dom. In the end, we can be assured, all human activity (mental and otherwise) that is equivalent to the execution of a well-defined set of rules can be replicated.[1] The chess program that finally defeats Gary Kasparov or his successor will no doubt be a *tour de force*.[2] But, I would argue, it demonstrates only how limited our unaided capacities really are.

How about non-logical or possibly "non-computable" mental processes. Recently, Penrose has argued that computers (Turing machines) cannot pass the Turing test because real brains performing real mental processes produce non-computable steps. And, in fact, non-logical processes, leaps (note: "leaps," never "steps") of invention or imagination pose somewhat of a problem for the gifted hacker. How does one write a set of rules that produce the unexpected, that compose the magnificent line:

> "Multitudinous seas incarnadine"

using two such grotesque, elephantine, English words?

To include such "illogical" steps into a machine's repertoire of possible responses requires a deeper understanding of how information is actually acquired, stored, and manipulated in the brain.

It has been obvious to many of us for quite a while that the brain is only marginally a computing system. It is no more designed for logic or reason than the hand is designed to play the piano. If it is designed at all, the design concerns survival and, in an ordered world, survival is enhanced by rapid (even if occasionally incorrect) decision-making. If I may be permitted to quote myself:

---

1 The games we play – as perhaps the scientific problems we attack – are chosen from all those possible to fit our actual mental and physical capabilities: difficult enough to be interesting, but not so difficult that we have no chance to succeed. Tic-tac-toe is not serious; two-dimensional chess taxes existing minds to the limit, while seven- or eight-dimensional chess is not usually attempted.

2 Kasparov was defeated by IBM's Big Blue in 1997. He is said to have complained that he was not even allowed a laptop to assist him.

The animal philosopher sophisticated enough to argue "the tiger ate my friend but that does not allow me to conclude that he might want to eat me" might then be a recent development whose survival depends on other less sophisticated animals who jump to conclusions.

In recent years progress has been made in understanding how information may be acquired, stored, and manipulated in a biological system. Although this work is in its infancy what are designated sometimes as neural networks and/or learning systems based on large numbers of processing units that mimic neurons and that can be trained with supervised or unsupervised learning methods have been extensively studied in the last generation. Systems now exist that recognize patterns, detect fraudulent behavior, and can be said to display, at least on a primitive level, features such as recognition, generalization, and association and which suggest some of the mental behavior associated with animal memory and learning.

Through years of painful education, somehow, our brain can achieve the ability to reason and execute rules as well as to associate. It seems clear that association is much easier and more natural than reasoning – the preserve of some of our more cerebral types.[3]

I say somehow because it is not yet all worked out (there are, of course, major areas of current research directed toward determining precisely how visual and other information processing are achieved) but, perhaps conceptually, it is there. We can already

---

[3] Among the glories of human intellectual achievement is, from the hazy associations that are the natural capability of the brain, just this creation and successful application to messy real-world situations of precise language (to make what is said depend on what was said before) and logical reasoning. In a *fin de siècle* cop out, a California update of dialectical materialism, Hegel come to Los Angeles, we witness a seasonal twist of intellectual fashion that heralds fuzzy logic (a sometimes useful engineering tool), recently become fuzzy thinking as more appropriate than Aristotelian logic to describe the less than sharp boundaries of the real world. Thus it is proposed that we convert the razor-sharp distinctions of the trained and athletic mind to the indistinct muttering and imprecise groping of the intellectual couch potato, that to the complexities of the world we add the charm of not knowing what we are talking about. *Sic transit . . .*

envisage how interacting associative and rule-based systems could function together to reason and associate (at least in simple situations) as we do. We expect that as our experience with rule-based and data-driven learning systems increases, we will more and more be able to reproduce (and finally surpass) human reasoning – or more precisely the output product of human mental activity and thus design a system that passes the Turing test.[4] [We might distinguish human from machine however by noting the time taken to respond to "What is $2\,579\,362 \times 1\,279\,854$?"]

Having designed such a system (its practical and commercial value aside), do we have a machine (a generalized Turing machine?) that thinks? In effect, this is the question the Turing test is designed to evade. An evasion due, in part, to an excess of positivism – a fear of and/or aversion to mentalism or assumptions about the internal workings of the mind that cannot be directly verified by experience. An evasion that, to my mind, is totally contrary to the nature and purpose of scientific thinking.

Successful science has given us just such machines (actual or conceptual) that work behind the actual events. The greatest include Newton's laws, Maxwell's equations, and Schrödinger's equation. Such entities as molecules and/or atoms were assumed to exist (an assumption that was vigorously contested in the nineteenth century with positivistic-type arguments) long before they were "seen." It is not necessarily the case that every element of the "behind the scenes" machinery can be directly observed. In quantum mechanics, for example, the wave function is not directly observable. The consequences, of this sometimes-invisible machinery can, however, be put into correspondence with experience. The essence of the positivist argument (as actually employed by Einstein and Heisenberg) is not that we cannot introduce

4 A more reasonable evolution – one that is already taking place – is the design of "reasoning and/or decision-making" systems that handle vast quantities of data with enormous rapidity, but applied to very specialized domains (e.g. airline reservations and seat assignments or fraud detection in credit card transactions).

entities that are not directly observable but, rather, that if an entity is not observable (e.g. absolute time in special relativity or simultaneous position and momentum in the quantum theory) it need not appear in theory.

Thus a satisfactory theory of mind not only is allowed but also, in my opinion, requires, the introduction of mental entities. We will be satisfied only when we see before us constructs that can have mental experience, when we see how they work, how they come about from more primitive entities such as neurons.

Further, in an ingenious construction (known as the "Chinese room"), John Searle has argued that purely algorithmic behavior, in this case an English speaker answering Chinese questions about a Chinese story with Chinese answers by applying purely syntactic rules for manipulating Chinese symbols does not necessarily imply understanding of either Chinese or the story.[5] Thus (although he agrees that computers might pass the Turing test) he argues that passing this test is not evidence of "thinking" or "understanding."

All of this can be summarized by saying that the Turing test is not sufficient. At best, it provides us a perfectly responding "black box" but no knowledge of how it all works – and that, in my opinion, is just the knowledge we want.

We could, of course, take the black box apart to investigate the sequence of instructions which gave us all of the right answers. It remains an open question whether this would necessarily shed any light on the phenomenon of mental activity.

What is more surprising, perhaps, is that for what is hardest to understand, the origin of the complex of mental experience: consciousness, awareness of ourselves, feeling, passing the Turing test is not necessary. Nothing in the perfect machine response gives us any indication of how such experience comes about. If I may quote myself again, we have absolutely no idea "how is it that a machine (for unless we accept a Cartesian dualism we are surely machines)

---

5 Searle, J. (1980). Minds, Brains and Programs, *Behavioral and Brain Sciences*, **3**(3), 417–457.

can feel?" Or in the words of the late philosopher Hans Jonas, "the capacity for feeling, which arose in all organisms, is the mother value of all."[6]

It is claimed by some (the proponents of "strong artificial intelligence") that mental qualities such as feeling, consciousness, and understanding emerge as a consequence of the execution of those algorithms that lead to the appropriate Turing responses. This may or may not be the case. But no one has given any indication of how it would come about.

Further, counter examples exist. Searle's "Chinese room," in my opinion, shows that correct algorithmic behavior can occur with no understanding, while every dog wagging its tail tells us that consciousness is there with little or no algorithmic behavior. It is much more plausible to believe that algorithmic behavior and consciousness are independent qualities.

Although some aspects of algorithmic behavior can clearly be achieved without conscious awareness, it remains possible that such awareness is required to attain the full power of our information processing abilities.

The deepest error and what may be most misleading is the attempt to equate the activity of the brain with a reasoning system. A more appropriate view is one that is attributed to Albert Szent-Györgyi:

The brain is not an organ of thinking but an organ of survival, like claws and fangs. It is made in such a way as to make us accept as truth that which is only advantage. It is an exceptional, almost pathological constitution one has, if one follows thoughts logically through, regardless of consequences. Such people make martyrs, apostles, or scientists, and mostly end on the stake, or in a chair, electric or academic.

For the understanding of such mental behavior as consciousness or feeling, it is thus clear that the Turing test is not only not necessary,

6 Jonas, H. (1992). The Burden and Blessing of Mortality, *The Hastings Center Report*, 22(1).

but is totally irrelevant. One can reason, or at least perform logical operations, without feeling: mechanical calculators do it all the time. Dogs and cats (even turtles, probably) feel but can't answer many questions.

In Turing's defense (if he needs a defense), we should note that such mental behavior was not what he had in mind. His concern was the output of human reasoning – not human feeling.

Let us then put this famous and ingenious criterion to rest and confront again the underlying problem. Can we understand the human mind (all of its components: reasoning, feeling, self-awareness) and its presumed origin in that biological organ, the brain? Could we, in the extreme, construct a machine that was conscious? (Whether a machine in this sense could be a Turing machine I will leave for others to answer.) What are the steps required so that a machine (algorithmic or not) can experience mental activity? The *non sequitur* "How would we know?" is an evasion. (How do we know anything?) Whether we can be sure if another creature and/or a machine are conscious is independent of the understanding of how it is that consciousness arises as a property of a very complex physical system. This, not reasoning power, is the profoundest mystery surrounding that biological entity, brain.

What has made this problem even more perplexing is the confusion of the various components of mental activity, the failure to distinguish what we believe we know – at least in principle – from what we do not yet understand at all. Although it is true that we are still far from understanding, for example, how the visual system processes and sorts information, such questions can be precisely formulated and it seems reasonable to believe that answers can be constructed from materials available. The same might be said for reasoning (logical and otherwise). But what is the source of our mental experience, our conscious awareness?

Mental awareness or consciousness themselves have many components, some easy to understand, some still incomprehensible: an on/off switch; memory in storage, memory in play; the distinction between ourselves and the external world (still the subject of vast philosophical cerebration) are, to my mind, easily understood.

But what, for example, is desire? No problem to program directives to pursue, functions to maximize or minimize, hot or cold to avoid, warm and cuddly to seek. What is the source of our self-awareness of the process, and possibly the mother of all: our ability to feel.

These questions are sufficiently difficult so that we have been subjected to the usual evasions – Cartesian dualisms: variations of homunculus proposals; solutions of one mystery by invoking another: consciousness arises in the quantum measurement process or where gravity meets quantum theory; total refusal to confront the issue: consciousness arises "somehow" when a machine executes the proper algorithmic processes; total retreat under the cover of positivist philosophy: how would we know if a machine was conscious... and so on.

We have heard such arguments before. They seem to be typical responses to the frustration of failure in attacking really difficult scientific problems. First try and fail. Follow this by proving a solution is impossible or irrelevant. Toy with the notion that a new law of nature is involved. Then, when the solution is found, complain that it is really trivial or (even better) that it was suggested in some obscure comment once made in a paper you published a long time ago.

On a personal note, we experienced all of this in the course of developing a theory of superconductivity – also a complex and subtle consequence of an interacting many component system, in this case the quantum mechanics of electrons in a metal. After the fact one, rather well known, physicist expressed his disappointment that "such a striking phenomenon as superconductivity [was] ... nothing more exciting than a footling small interaction between electrons and lattice vibrations," thus missing the point in operatic style.

It could turn out that we must invoke a new "law of nature": pour the conscious substance into the machine, a position not unfriendly to a common view of mankind in its emerging years. But the conservative scientific position is to attempt to construct this seemingly new and surely very subtle property from the materials available – those given to us by physicists, chemists, and biologists (as has been done many times before: celestial from earthly material, organic from inorganic substances, the concept of temperature from the motions of molecules, and so on). If this cannot be done (perhaps one could be patient enough to give us a couple of years to try) then we will genuinely have made one of the profoundest discoveries in the history of thought, consequences of which would shape and alter our conception of ourselves in the deepest way.

To paraphrase Shannon: "Can machines feel?" "Sure they can. We're machines. We feel." But are we? If so, how?

16

Mind as Machine: Will We Rubbish Human Experience?

Anxiety about robots and other forms of artificial intelligence precedes our ability to actually construct such entities. But does the ability to create thinking machines invalidate human experience? Will robots replace or supplant humans? Not if we can help it.

This essay is based on a talk originally given at the conference "Neural Networks and the Mind, Interdisciplinary Conference on Culture and Technology in the Twenty-First Century: Brain Research – An Intervention in Culture," in Dusseldorf, Germany, 1993.

Recently, Hans Moravec discussed a transition from humans to what he calls a "Universal Robot."[1] The prospect that such a transition might be carried out inspires fear and raises many doubts. In this lecture, we analyze some of the problems that lie in the path of constructing robots or machines that think. In particular, we discuss the relation of Turing's famous test to a theory of mind and, drawing on the wisdom of Lessing and Goethe, explore possible implications for the meaning and worth of human experience.

In the program for the 1993 conference, "Neural Networks and the Mind," in Dusseldorf, Germany we read:

A scientific revolution is in the air. Over the last 20 years there have been more scientific discoveries made concerning the human brain than in all

---

1 Moravec, H. (1999). The Universal Robot, in *Ars Electronica: Facing the Future*, T. Druckrey (ed.), Cambridge: MIT Press, pp. 116–123.

of the previous 200 years. Is brain research on the point of providing scientific explanations for our thinking and feeling processes, for our consciousness and our personal identity? A branch of science which has seen a rapid development over the last ten years is addressing the very centre of human existence. Is the brain the seat of the soul? Is it in the brain that our ego resides?

Last year Hans Moravec presented a vision of the future in which human beings would be replaced by what he called "Universal Robots." Neither the vision nor even the name is new. At the turn of the century, Karol Kapok fascinated and repelled audiences with his play *RUR* (*Rossum's Universal Robots*). To some, this is a brilliant vision of the future. To others, it is a variation of apocalypse.

We do seem to be on the threshold of deep new understanding of the brain. Such questions as the physiological basis for learning and for memory storage may soon be answered. But how would knowledge of events on the molecular and cellular level relate to human thought? Is complex mental behavior a large system property of the enormous numbers of units that are the brain? And if we understand the answers to such questions, what will it mean for us?

Perhaps part of the answer lies in a work that should be familiar to you. You may even have memorized the first few lines:

Habe nun, ach! Philosophie
Juristerei und Medizin,
Und leider auch Theologie
Durchaus studiert, mit heißem Bemühn.
Da steh'ich nun, ich aimer Tor,
Und bin so klug als wie zuvor![2]

---

2 "Look at me, Years wasted grinding through philosophy, Slaving over medicine and law, Learning everything…Now here I am with all this lore, Poor fool no wiser than before."

In this article, I analyze some of the problems that lie in the path of constructing robots or machines that think. In particular, I discuss the problems involved in constructing a theory of mind and, drawing on the wisdom of Lessing and Goethe, explore possible implications for the meaning and worth of human experience.

In recent years progress has been made in understanding how information may be acquired, stored, and manipulated in a biological system. Although this work is in its infancy, what are designated sometimes as neural network and/or learning systems, based on large numbers of processing units that mimic neurons and that can be trained with supervised or unsupervised learning methods, have been extensively studied in the last generation. Systems now exist that recognize patterns, detect fraudulent behavior, and can be said to display, at least on a primitive level, features such as recognition, generalization, and association and which suggest some of the mental behavior associated with animal memory and learning.

Having designed such a system (its practical and commercial value aside), do we have a machine that thinks? It is claimed by some (the proponents of "strong artificial intelligence") that mental qualities such as feeling, consciousness, and understanding emerge as a consequence of the execution of those algorithms that lead to the appropriate Turing responses. This may or may not be the case. But no one has given any indication of how it would come about.

The problem becomes: Can we understand the human mind (all of its components: reasoning, feeling, self-awareness) and its presumed origin in that biological organ, the brain? Could we, in the extreme, construct a machine that was conscious? What are the steps required so that a machine (algorithmic or not) can experience mental activity? The conservative scientific position is to attempt to construct this seemingly new and surely very subtle property from the materials available – those given to us by physicists, chemists, and biologists (as has been done many times before:

celestial from earthly material, organic from inorganic substances, the concept of temperature from the motions of molecules, and so on). Or, perhaps, as Moravec describes, somehow transfer the contents of the brain to animate the robot machine.

A distinction must be made between a new assumption, an entirely new entity (the equivalent of the addition of Euclid's fifth or parallel axiom to the first four that distinguishes between Euclidean and non-Euclidean geometries) and an unexpected and non-inevitable construction from the materials available (e.g. living creatures from chemicals or novels from letters). These latter are often highly dependent on initial conditions. Even temperature, a seemingly straightforward construction from kinetic theory and statistical mechanics, requires equilibrium systems.

The prospect that such a program could be carried out elicits occasional paranoid reactions: cries of reductionism or, as expressed by John Lucas of Oxford, "the rubbishing of human experience."[3]

As for reductionism, I have always been somewhat mystified as to what the fuss is about. Scientists (as mentioned above) have been constructing seemingly new and elevated entities from base material as part of their daily exertions since Thales showed us the way. When such a construction cannot be made, something new must be added. There seems to be no shortage of voices advising us to add that something new before we have had a reasonable chance to construct from the old. My advice is patience, along with the reminder that our brain is programmed to jump to conclusions.

The rubbishing of human experience is a greater concern since no small number of the *nouvelle vague* computer and robotic types seem only too happy to do just that. To me this is a reflection of the joy they have experienced and the wisdom they have gained in

3 Lucas, J. R. (1963). Minds, Machines and Gödel, in *The Modeling of Mind*, K. M. Sayre and F. J. Crosson (eds.), South Bend, IN: Notre Dame University Press, pp. 269–270.

their passage through this vale of tears. The value we place on our own experience is something we determine ourselves and would never (I hope) forfeit to any machine or in fact (except to a limited extent) to anyone else. This value is completely independent and totally unaffected by any "reductionist" explanation of how our mental activity comes about. No more than a detailed knowledge of the chemistry of digestion affects our appreciation of the product of a great chef or the bouquet of a fine wine.

Well then how, from materials made available to us by unrepentant reductionists: electrons, protons, atoms, molecules, DNA, RNA, receptors, enzymes, proteins, membranes, neurons, axons, dendrites, synapses...can we construct an entity that has mental experience?

We believe that there is a reasonable evolutionary sequence leading from sunshine, lightning, and a reducing atmosphere to molecules, the primitive protein soup and to more and more complex structures. Even the simplest cells show reflex-like, chemically directed responses to various stimuli (aversion or attraction). As has been said, "protoplasm is irritable." (But how does it come to feel irritable?) It seems reasonable to believe that there is advantage for organisms that can communicate from one end to the other, that electrical communication is a very efficient way of doing this and that excitable membranes provide the means.

There is further advantage in the innovation of a nervous system that is plastic. (The transmitter-driven synapse provides an excellent option.) For the animal can now learn and store memories of past experience. This animal can adapt to environmental changes in less than evolutionary time.

And there is surely advantage (most of the time) in exercising the option of mental experience – feelings of pleasure and pain, awareness of individuality, instilling a directive to "jump to conclusions" to "accept as truth that which is only advantage" as a means of producing life preserving behavior in complex and somewhat unpredictable real-world situations.

All of this can at least be sketched. We can guess (at least conceptually) how from primitive feelings such as pleasure and pain more complex mental states could be constructed. But how the essential primitive – feeling – arose out of materials such as reflex reactions to hot and cold, how, somewhere in the distinction between those events that produce physical reactions in ourselves and those that do not, in the interplay of present sensory input with memory of past experience, our self-awareness, our mental experience, our consciousness arose, remains a deep mystery.

And if this mystery is solved, as we expect it will be, what will it mean? Are we doomed to be replaced by Moravec's robots or by some genetically engineered equivalent? If we can do successful genetic engineering, do we want to replicate everything, lower back pain and all?

Today we can replace knees, hips, kidneys, lungs, and even hearts. One day we may be able to transfer the contents of our mind into another body (as I suggested in an unpublished novel twenty years ago) or into a replacement machine with far superior memory capacity and processing power (as proposed by Hans Moravec).

No one, these days, suffers philosophical anguish over the prospect of living with a mechanical hip, another person's cornea, or even another person's heart. (How would Aristotle, who classified heart as the seat of intellect and reserved for brain the function of cooling system – which by the way it is, 600 calories a day being dissipated there – have felt?) How would we react to living with another person's brain? Who is the person that would be living?

This most recent human voyage, intriguing, tempting, but troubling is already en route. How far will we go? Are we on the brink of Ponce de Leon's fountain of youth or the endless prolongation of old age? If we can replace everything, do we have a new machine or a continuation of the old? As is often the case in life and in theatre, only the scenery is new. The argument has been with us for quite a while and the poet has already addressed it.

Goethe's masterpiece begins with Faust old and discouraged:

> ... *Da steh'ich nun, ich armer Tor*
> *Und bin so klug als wie zuvor!*"[4]

Faust knew everything that could be known and was not content. Through Mephistopheles he achieves "All that is given to humanity to experience..." In the course of this he becomes further and further separated from Mephistopheles. And in the end he is saved as though in Lessing's words:

God has not given Man the noblest of impulses only to make him eternally miserable.

Somehow "the force that always tries to do evil" has once more done good.

In all of the poem's richness and diversity, Faust's gradual but inexorable separation from Mephistopheles is particularly intriguing and suggestive. Faust progresses, in the course of his adventures from initial exultation, to try everything to grasp "all that is given to humanity to experience...the heights and depths of it...," to his own esthetic criteria elaborately constructed from classical and romantic models, to his final desire, to create a new Utopian society (to the utter contempt of Mephistopheles) and it is this he thinks he is doing when he is moved to utter the fatal words to the passing moment: "Oh stay a while, you are beautiful..."

It is just such experience, Goethe is saying, the experience of each of us:

> *del camin de nostra vita,*[5]

experience refined by intellect, reason, and discernment that enriches and ennobles. To achieve this state of exaltation we must feel as well as know; experience must be active not passive (television is not a replacement for being there); the great decisions of life

---

4 "Now here I am with all this lore, / Poor fool no wiser than before."
5 "The journey of our life."

are which experience to choose (since we cannot choose all; only Faust was given that option); unpleasant is sometimes as valuable as pleasant.

Thus, we conclude that physical replacements we make to our bodies that enhance or continue our ability to experience and continue the memory of that particular sequence of experience – mental as well as physical – that makes an individual pose no threat. It is the replacement of the feeling and reasoning entity that we are by an entity that only reasons and/or knows that would be disturbing because it eliminates the maturation of spirit through experience and its refinement through intellect and reflection by a non-feeling machine.

My personal choice would be that we employ Universal Robots to do the dishes, clean the floors, and dispose of nuclear waste.

17

Memories and Memory: A Physicist's Approach to the Brain

It is remarkable how little we need to know about ourselves in order to survive and reproduce. Humans are probably the only animals that have made conjectures about what goes on inside the body. And even then, it has taken a long time to arrive at a clue.

This essay is based on an article originally published in the International Journal of Physics, A, *15(26), in 2000.*

It was less than four hundred years ago that Harvey taught us that the heart was a pump. The great Aristotle had conjectured that the heart was the seat of intellect, reserving for the brain the function of a cooling system, which (in addition to some other functions of more than minor interest) it is. In dealing with questions about our own nature we have to survive ideology – endless discussions about organic versus inorganic, vital versus non-vital, living versus non-living. These hotly disputed questions quietly fade away when the cool light of patient investigation and hard thought finally provide illumination. Today we are engaged in the quest of understanding our brain. When we have finally worked out the details, this remarkable organ, for all of its lofty pretensions – seat of intellect, home of the soul – will very likely join other remarkable pieces of biological machinery. Remarkable, certainly, but not mysterious or possessed of any supernatural qualities.

How then do we go about trying to understand a system as complex as the brain? Obviously we cannot just make observations. The number of possible observations is substantially larger than the available number of scientist-hours, even projecting several centuries into the future; the result might be a listing of facts that would be of little use. A theory or a point of view is essential. These shape the direction of analysis as well as guide us toward what we believe are the relevant observations. Theory provides a framework within which questions become relevant.

The good materialist fervently believes that everything in our world, from vacuum polarization contributions to the electron's magnetic moment, to chemical reactions, mental states, and social behavior, can be constructed from such objects as protons, electrons, quarks, strings, or branes: objects that obey those few remarkably concise rules known as the "laws of physics." Why then don't these "laws" explain everything? We might as well ask why they don't explain *Hamlet*. No small number of stage works (good, bad, or indifferent, as well as a great deal of gibberish) can and have been constructed using sequences of letters, spaces, and punctuation made of quarks, branes, etc. that happily obey the "laws of physics." Whatever it is that distinguishes *Hamlet* from, let us say, *The Importance of Being Earnest* is not easily gleaned from these "laws" since they seem perfectly content with both.

Such an argument is, perhaps, appropriate for theatre; but does it apply to science? I believe it does. Consider Darwin's theory: Evolution dominated by competition for limited resources and natural selection is considered by most (excluding, perhaps, the state of Kansas) to be the major principle governing the development of those species that are here on earth. But though Darwin's evolution is consistent with the "laws of physics," it is not an inevitable consequence. (Consider, for example, an environment with relatively unlimited resources dominated by recurring catastrophic events that randomly destroy most living creatures.)

Explanation for complex systems, from stage works, human behavior, the evolution of the species, to the properties of systems of neurons, requires choosing among those many objects or rules consistent with the laws of the physics, to distinguish what is from what might be.

In this article, I would like to describe the path I took in attempting to understand important aspects of one complex biological system: the human brain.

The first question one asks one's self before embarking on such a hazardous intellectual journey, surely, is: Can it be done? The second, perhaps, is: Can it be done in our lifetime (being too far ahead of one's time is not wise either in science or in business); a third, reasonably, is: Can I contribute?

In the early 1970s it was generally believed that the special properties of nervous systems, in particular the brain, are the results of interactions between neurons, specialized cells that transmit information chemically and electrically and are linked to one another in networks of incredible complexity. Although the properties of individual neurons as well as the connection, for example, between neuron and muscle activity were more or less understood, higher-level "mental" activity such as the means and place of memory storage seemed completely mysterious.

Here then was a possible opportunity. Although the brain is certainly a very complex piece of biological machinery, the functions it performs seemed to require organizing principles whose elucidation might require the talents of, and be amenable to, analysis by a theorist.

My first effort in this area was to attempt to construct a network of neurons that would display some of the qualitative features associated with what I called animal memory. Initial success was the seduction that lured me further and further into this exotic domain. It is this journey I will describe.

Neurons are very complex cells. The result of large number of simplifications leads to a relation between neuron inputs and output as in Figure 17.1a; a simple network of such neurons might look as shown in Figure 17.1b.

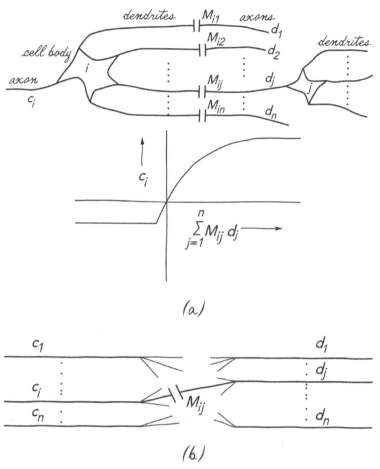

Fig. 17.1 Neurons are very complex cells. (a) The input from the jth incoming cell is d_j. M_{ij} is the matrix of "ideal" synaptic junctions and c_i is the output of the ith cell. (b) A simple network of neurons. M_{ij} is the "ideal" synapse connecting the jth input cell with the ith output cell.

Suppose that an "external auditory event," such as a particular tone, is eventually mapped into n neurons (presumably in auditory cortex).

$$d = (d_1 \cdots d_n) \longleftarrow \text{tone} \qquad (17.1)$$

Suppose, further, that in the experience of the animal, this tone is accompanied by or followed by a visual event, the sight of food (presumably in some region of visual cortex) that triggers salivation.

$$\text{salivation} \longleftarrow c = (c_1 \ldots c_n) \longleftarrow \text{sight of food} \quad (17.2)$$

It is known that with a sufficient number of such experiences the animal comes to "associate" the tone with the sight of food and eventually begins to salivate upon hearing the tone:

$$\text{salivation} \longleftarrow c < \text{------} d \longleftarrow \text{tone} \quad (17.3)$$

We might have n such associations summarized by the mapping:

$$
\begin{aligned}
c^l &< \text{-----------} d^l \\
c^k &< \text{-----------} d^k \\
c^n &< \text{-----------} d^n
\end{aligned}
\quad (17.4)
$$

Could such "associative" and "content addressable" mappings be constructed in simple neural networks? If we look more closely and ask what is happening biologically, what is happening at a particular synaptic junction, we see that the required strength of the ijth synapse, M_{ij}, is a sum over all associations $(l \ldots k \ldots n)$ of the product of the output of the ith neuron and the input from the jth neuron (see Figure 17.2a).

Such a synaptic strength could result if synaptic modification (or learning) follows the famous Hebbian rule that requires that information of the summed postsynaptic potential be propagated back from the cell body to individual synapses. (See Figure 2b).[1] (It is immediately clear that Hebbian learning can be only part of the story since synapses would grow in strength without bound. Thus one early question was: How could such learning be stabilized?)

---

1 "Back-spiking" that could carry this information has recently been observed, and associated with changes in synaptic strength.

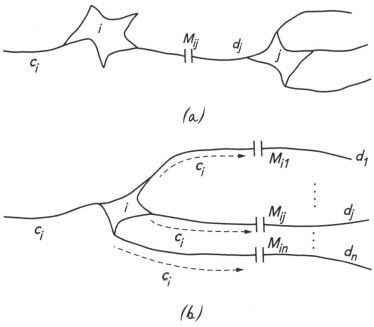

Fig. 17.2 (a) The axon of the jth incoming neuron is connected to the dendrites of the ith outgoing neuron by many actual synapses. Their net effect is given by the strength of the single "ideal" synapse, M_{ij}. (b) In order for the information required for Hebbian synaptic modification to be available locally at each cell's synapses, M_{i1}, ..., M_{in}, the integrated potential, c_j, must be propagated backwards (in a direction opposite to the usual information flow) from the cell body to each of the synapses.

Although such conjectures seemed attractive, in the early 1970s there was little if any evidence for synaptic modification of any kind. The appearance of vaporware was difficult to avoid. Many ideas, some possibly interesting, but with no real connection to the world in which we happen to live, gave a limited contribution to a field, if one could call it a field, that was and is plagued with excessive mathematical and philosophical wheel-spinning.

It seemed essential to me that theory be made sufficiently concrete so that it could be confronted by experimental results. At the time theory was a somewhat novel idea for biologists: plausibly so since, for the most part, that specialty, the spinning out of consequences of ideas in long and complex arguments, while accepted (although occasionally the subject of some ridicule) in the community of physicists, had not really been required in biology. There, the connection between idea and experiment was straightforward enough so that every self-respecting experimentalist insisted on doing it himself. This skepticism was also justified, in my opinion, since with a few striking exceptions, many previous so-called theoretical attempts were totally removed from reality. Physicists, in particular, displayed an arrogance in talking to biologists that was not designed to inspire friendly relations. I recall a presentation at a conference in which an eminent physicist said, in effect, *"Here is the Schrödinger equation. Here we have 10^{23} electrons and ions subject to electrical forces. One of the consequences is life."*

We thus began a series of attempts to make specific connections between fundamental ideas of synaptic modification and testable experiments in actual animals, to produce a theoretical structure concrete enough so that one would know precisely what the assumptions were, and so that one could see one's way through the arguments and know exactly which conclusions followed from which assumptions. The primary object is not to be right (although that certainly is one of the hopes), it is to be crystal clear so, to paraphrase Galileo, "One knows what follows from what one has said before." His teachers of mathematics taught him this method.

We chose visual cortex because of the large number of experiments that had been done in that region of the brain. At the time, in addition to the work of Hubel and Wiesel, there was a large body of (very controversial) experimental work suggesting that the response properties of cells in visual cortex depended on the visual experience of the animal. This indicated to us that one might be observing experience-dependent cellular changes, analysis of which could reveal the systematics of synaptic modification. The collaboration between experimentalists and theorists that resulted has continued until the present. The primary questions at that time seemed to be: Can we find any evidence for synaptic modification? If so, what is its form? Further, what is its cellular and molecular basis, thus the cellular and molecular basis for learning and memory storage?

One path toward making contact with experiment began with the observation that in most situations, there would be confusion among the outputs. To better separate incoming events, we can try to form selective neurons (those that respond more to the mapping of one external event than another).

Such selectivity is relatively common in the nervous system. Hubel and Wiesel observed edge detectors in area 17 (V1) of visual cortex of kittens. Rolls and colleagues observed face detectors in inferotemporal cortex of monkeys. (Humans with lesions in inferotemporal cortex suffer from a rare disorder called prosopagnosia, a severe disturbance in the ability to recognize familiar faces.) Further the development of such selectivity in visual cortex had been shown to depend on the animal's visual experience.

Hebbian synaptic modification yields no selectivity and needs stabilization. To construct a mapping by a learning process that proceeds without instruction on the presentation of overlapping inputs, we introduced a form of synaptic modification that combined Hebbian with anti-Hebbian regions. Then, allowing the crossover point between these two regions to vary with cell activity led to stabilization. The resulting form has become known as BCM synaptic modification (see Figure 17.3).[2] In general one can test and/or distinguish between theories by comparing predicted consequences of theory with experiment, or by attempting, more or less directly, to experimentally verify the underlying assumptions.

2 Bienenstock, E. L., Cooper, L. N, and Munro, P. W. (1982). Theory for the Development of Neuron Selectivity: Orientation Specificity and Binocular Interaction in Visual Cortex, *Journal of Neuroscience*, 2(32), 32–48, and Intrator, N. and Cooper, L. N (1992). Objective Function Formulation of the BCM Theory of Visual Cortical Plasticity: Statistical Connections, Stability Conditions, *Neural Networks*, 5(1), 3–17.

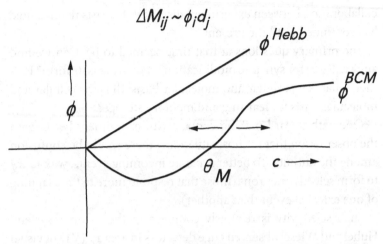

Fig. 17.3 BCM synaptic modification. The BCM synaptic modification function is contrasted with Hebbian modification function. Among the important results are: (1) fixed points depend on the environment; (2) only selective fixed points are stable.

Consequences of BCM synaptic modification have been shown by analysis and in many simulations to be in agreement with experimental observations on the receptive field properties of visual cortical cells for animals reared in normal and various deprived environments (see Figure 17.4a).

These might be characterized as post-dictions since the deprived rearing experiments were known for the most part (with the exception of the reverse suture result) before the BCM theory was constructed. Although there is no logical distinction between post-diction and prediction (all theorems must be in agreement with observation) it is psychologically satisfying to have a clear prediction that is confirmed by experiment. One such prediction, that has recently been experimentally tested, is described below.

A dramatic example of experience-dependent receptive field plasticity is the shift in ocular dominance in visual cortex that results from briefly depriving one eye of vision. Such deprivation leads to a very rapid disconnection of the deprived eye from cells

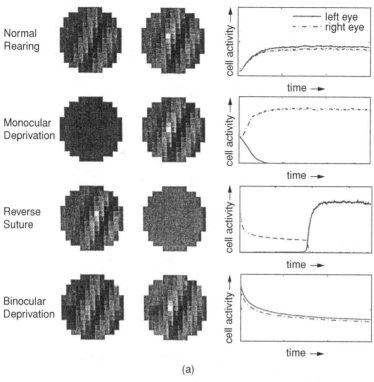

(a)

Fig. 17.4 (a) Simulations of the development of cortical receptive fields using BCM synaptic modification. Left: Final receptive fields and synaptic weight configurations. Each pixel represents a point in space over the retina, where white and black correspond to strong and weak synaptic strengths, respectively, from that retinal input. Right: Maximum response to oriented stimuli, as a function of time. Simulations from top to bottom are as follows. Normal rearing: both eyes presented with natural scenes. Monocular deprivation: following normal rearing, the left eye is presented with noise and the right with natural scenes. Reverse suture: following monocular deprivation, the eye presented with noise is now presented with natural scenes, and the other eye with noise. Binocular deprivation: following normal rearing, both eyes are presented with noise. It is important to note that if the binocular deprivation simulation is run long enough, selectivity will be lost. These simulation results are in agreement with experimental observations, (b) The effect of deprived eye activity on the disconnection of the closed eye in monocular deprivation with BCM synaptic modification. Shown are the results of monocular

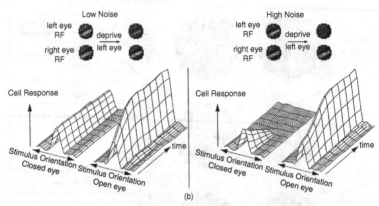

(b)

Fig. 17.4 (cont.) deprivation starting from the binocular state. Left and right receptive fields (above), before and after depriving the left eye. Each pixel represents a point in space over the retina, where white and black correspond to strong and weak synaptic strengths, respectively, from that retinal input. The responses of the cell to oriented sine gratings (lower plots), as a function of time during deprivation in a low noise environment (lower left) and a high noise environment (lower right). Note that with higher noise the disconnection is more rapid (see equation on page 9).[3] This is in agreement with the experimental results of Rittenhouse et al.[4]

in visual cortex so that stimulation through the deprived eye no longer drives the cortical cells. Analysis of the BCM modification equations leads to the result that as the noise from the closed eye increases, so will the disconnection rate of the synapses from the closed eye. Thus, one has the non-intuitive result that higher levels of noise lead to a faster disconnection of the closed eye. This is an experimentally testable prediction of the BCM theory, and one that distinguishes this theory from others (see Figure 17.4b). A recently completed experiment, inspired by the above analysis,[5] is consistent with the BCM prediction that deprived eye

3 Blais, B. S., Shouval, H. Z., and Cooper, L. N (1999). The Role of Presynaptic Activity in Monocular Deprivation: Comparison of Homosynaptic and Heterosynaptic Mechanisms, *Proceedings of the National Academy of Sciences USA*, **96**(3), 1083–1087.
4 Rittenhouse, C. D., Shouval, H. Z., Paradiso, M. A., and Bear, M. F. (1999). Monocular Deprivation Induces Homosynaptic Long-term Depression in Visual Cortex, *Nature*, **397**, 347–350.
5 Ibid.

connections in the visual cortex are more rapidly weakened with increasing noise level.[6] Experiments to test the underlying postulates of the BCM theory have been performed at Brown in the laboratory of Mark Bear as well as elsewhere. The central assumptions of BCM synaptic modification are that (1) active synapses are bidirectionally modifiable, (2) the sign and magnitude of the modifications depends on the integrated level of postsynaptic response, and (3) the synaptic depression–potentiation crossover point (θ_m) varies as a function of the history of postsynaptic cellular activity (see Figure 17.3). Much experimental work over the past decade has been devoted to determining if these assumptions are valid at excitatory glutamatergic synapses in the cerebral cortex.

The basic shape of the BCM synaptic modification function was first confirmed by Dudek and Bear (see Figure 17.5a) in a region of the brain called hippocampus.[7] Kirkwood, Rioult, and Bear showed that the result was the same in visual cortex.[8] Since then these findings have been confirmed in many different regions of neocortex in many species in both young and old animals. Of particular interest are recent data showing that the same principles of synaptic plasticity apply in the human inferotemporal cortex, a region believed to be a repository of visual memories.[9] Together, the data support the idea that very similar principles guide synaptic plasticity in many species in widely different regions of the brain.

To stabilize the system, the modification threshold, θ_m, must vary according to the history of postsynaptic cortical activity. An

6 Recent work has complicated this analysis but the final result (at this moment) is that BCM is in agreement with experiment.
7 Dudek, S. M. and Bear, M. F. (1992). Homosynaptic Long-term Depression in Area CA1 of Hippocampus and Effects of N-methyl-D-aspartate Receptor Blockade, *Proceedings of the National Academy of Sciences USA*, **89**(10), 4363–4367.
8 Kirkwood, A., Rioult, M. G., and Bear, M. F. (1996). Experience-dependent Modification of Synaptic Plasticity in Visual Cortex, *Nature*, **381**, 526–528.
9 Chen, W. R. *et al.* (1996). Long-term Modifications of Synaptic Efficacy in the Human Inferior and Middle Temporal Cortex, *Proceedings of the National Academy of Sciences USA*, **93**(15), 8011–8015.

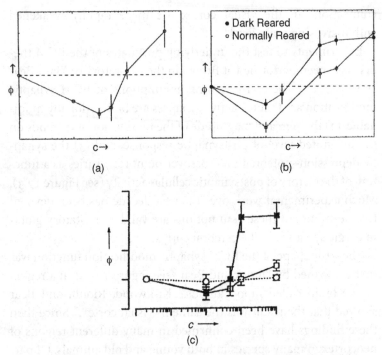

Fig. 17.5 (a) The synaptic modification function as constructed from the experimental results of Dudek and Bear in rat hippocampus. These results have been replicated in many parts of the brain, in young and old animals, and in many species – including humans. (b) The movement of the modification threshold as constructed from the experimental results of Kirkwood et al. on dark reared rats (filled circles) and those with normal visual experience (open circles). This activity dependent shift is consistent with the BCM postulate of the moving modification threshold. (c) The movement of the modification threshold as constructed from the experimental results of Tang et al. using genetically engineered mice and those of Quinlan et al. linking cellular activity with the ratio of two distinct subunits of the NMDA receptor.

experimental test of this hypothesis was first reported by Kirkwood et al.[10] They compared the synaptic modification function in the visual cortex of normal animals with that in the visual cortex of

10 Kirkwood, A., Rioult, M. G., and Bear, M. F. (1996).

animals reared in complete darkness and found a shift of this function in accordance with the theoretical postulate (Figure 17.5b). A very different experiment using genetically altered mice has been interpreted as confirming this result.[11] Also, recently published results of Tang *et al.*[12] again using genetically altered mice (a different alteration) taken together with results of Quinlan *et al.*[13] that link cellular activity with the ratio of two distinct subunits of the cortical NMDA receptor support the idea that θ_m is set according to the activation history of the cell (see Figure 17.5c).

On the basis of results, such as those sketched above, we conclude, with a certain optimism, that comparison of the consequences of BCM synaptic modification with experiment is satisfactory and further that the postulates that underlie the theory are consistent with experiment. It has become accepted that the synapses modify in a manner described, at least in a first approximation, in Figure 17.3.

Much effort is now devoted to investigations of the cellular and molecular mechanisms that underlie these synaptic changes. Extensive experimental work has revealed the dependence of synaptic modification on the influx of calcium into cells when accompanied by activation of the synapses. (This is presumed to be the cellular event that correlates active synapses with the integrated postsynaptic response.) It is generally believed that various receptors in the postsynaptic membrane play critical roles and that the modification of some of them through such processes as phosphorylation or changes of subunit ratios are responsible for

11 Mayford, M., Wang, J., Kandel, E., and O'Dell, T. (1995). CaMKII Regulates the Frequency-Response Function of Hippocampal Synapses for the Production of both LTD and LTP, *Cell*, **81**(1), 891–904.

12 Tang, Y. P., Shimizu, E., Dube, G. R., Rampon, C., Kerchner, G. A., Zhuo, M., Liu, G., and Tsien, J. Z. (1999). Genetic Enhancement of Learning and Memory in Mice, *Nature*, **401**, 63–69.

13 Quinlan, E. M., Olstein, D. H., and Bear, M. F. (1999). Bidirectional, Experience-dependent Regulation of NMDA Receptor Subunit Composition in Rat Visual Cortex during Postnatal Development, *Proceedings of the National Academy of Sciences USA*, **96**(22), 12 876–12 880.

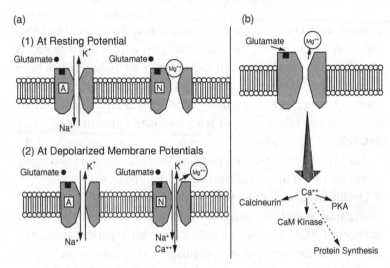

Fig. 17.6 (a) Postsynaptic response to glutamate. Among the receptors on the postsynaptic membrane that are activated by the neurotransmitter glutamate are two of particular importance: the AMPA receptor (A) and the NMDA receptor (N). The AMPA receptor responds to glutamate by opening its channel and allowing sodium and potassium ions to pass through. Even in the presence of glutamate, the NMDA receptor is blocked by Mg^+, its channel is opened only with a substantial depolarization of the postsynaptic membrane. Thus a coincidence of glutamate and sufficient depolarization are required to open the NMDA channel. When this channel is opened, it allows Ca^{++} to flow into the postsynaptic dendrite. (b) Modulation of the NMDA receptor effectiveness. The Ca^{++} that enters the cell is thought to initiate a sequence of molecular events that (depending on the amount of calcium that enters) result in an increase or decrease of synaptic strength. (This change in synaptic strength has been associated with the phosphorylation or dephosphorylation of sites on the AMPA receptor.) It has recently been shown that alterations in the ratios of subunits that make up the NMDA receptor change its response properties in a manner consistent with the BCM moving threshold postulate.

synaptic changes. Some of these, therefore, are thought to provide part of the molecular subtrate for memory storage (see Figure 17.6). The detailed molecular and genetic processes that are responsible for these changes as well as those that control the transfer of short or intermediate memories into long-term storage is now a subject of intense interest.

Given the level of skepticism displayed when ideas such as synaptic modification were discussed twenty-five years ago, I think that it is reasonable to say that we have made more progress than is generally appreciated. Our initial aim to build a theoretical structure relevant to a fundamental brain process that was sufficiently concrete so that it could be tested by experiment has been accomplished. It is particularly gratifying that theory has inspired experiments which, in addition to confirming the various postulates and predictions have led to the discovery of new phenomena. These have also allowed us to refine the hypothesis on which the theory was founded: to make sensible the introduction of complications to the originally very simplified assumptions that, hopefully, will lead more and more realistic descriptions of the processes of learning and memory storage.

Permit me then to project into the future with a certain optimism. Assume that the fundamental mechanisms are understood, that the neural (cellular and molecular) correlates of mental states (as Francis Crick is seeking) are known. Do we then understand the human mind and its presumed origin in that biological organ, the brain? Will we understand the origin of our feeling, awareness, or consciousness? This question is addressed in the next essay.

18

On the Problem of Consciousness

Can we understand consciousness? Can consciousness be constructed from ordinary materials? The implications would be monumental if it could, but would be no less so if it could not.

This essay is based on an article originally published in Neural Networks, *20(9), in 2007.*

Toward the end of his wonderfully productive life, Francis Crick engaged in a search for neural correlates of mental states. At least a glimmer of possible correlates is provided by techniques such as magnetic resonance imaging (MRI); these have been extraordinarily effective in determining regions of the brain involved in various mental activities. An additional glimmer is provided by the remarkable progress that has been made in elucidating the cellular and molecular basis for learning and memory storage. Although these are suggestive of the neural correlates for which Crick was searching and although many words have been expended on possible substrates for mental states, the origin and nature of consciousness, our awareness of ourselves, remains a complete mystery.

The mystery is sufficiently vexing to provoke occasional claims that the problem is not soluble using ordinary scientific methods. I have heard it argued, for example, that consciousness is an epiphenomenon (secondary phenomenon). (I'm not sure what is intended by this argument, except to suggest that consciousness is not really there and so doesn't have to be explained.) I would like to address some of these, let us call them philosophical, arguments and then state in a very simple manner what, in my opinion, is the heart of the problem.

First, and I believe most would agree, consciousness is not only there, it is a primary sense datum; it is one of the most immediate and certain of our experiences. So we must certainly try to understand its origin and nature.

One difficulty is that of obtaining an objective understanding of what seems to be intrinsically subjective. Can we understand subjective phenomena such as our self-awareness, our experience of color or touch (what are sometimes called *qualia*) in an objective manner? I believe the answer is yes. As has been proposed by John Searle, consciousness is ontologically (in its nature) subjective, but epistemologically objective. That is to say, knowledge about the nature of the subjective phenomenon consciousness can be objective.

Another difficulty is due to confusion about what is called scientific reduction. An underlying assumption science would like to make is that everything in the world is composed of materials given to us by generous physicists: quarks, electrons, protons, etc. (We might call this materialism.) So far, we have not needed to invoke a Cartesian-like dualism in which a mysterious, non-material, substance is poured into the material body.

This implies that we must explain or construct all that exists, everything that we experience, what we feel, our consciousness, our awareness of ourselves and others, from electrons, protons, atoms, molecules, and cells. (It does not imply, as I have argued elsewhere, that the "laws of physics" explain everything.[1])

Yet another difficulty is due, in part, to an excess of positivism – a fear of and/or aversion to mentalism or assumptions about the internal workings of the mind that cannot be directly verified. But successful science has given us just such structures (actual or conceptual) that work behind events in the world. The greatest include

---

1 See Chapter 17, "Memories and Memory: A Physicist's Approach to the Brain."

Newton's laws, Maxwell's equations, and Schrödinger's equation. The essence of the positivist argument as actually employed by Einstein and Heisenberg is not that we cannot introduce entities that are not directly observable, but rather that if an entity is not observable (e.g. absolute time in special relativity or simultaneous position and momentum in quantum theory) it *need* not appear in the theory.

Thus a theory of mind is not only allowed, but, in my opinion, requires the introduction of mental entities. We will be satisfied only when we see before us constructs that have mental experience, when we see how they work and how they come about from ordinary material. Reduced to its essence, the problem, as I see it, is to construct from components such as neurons the simplest entity that performs the most primitive conscious act. Could we, in the extreme, build a machine that is conscious?

We must understand how consciousness arises out of a system of neurons that themselves (presumably) are not conscious. This is the profoundest mystery surrounding the biological entity, brain.

The problem, in my opinion, is beautifully framed in a single sentence given to us by Santayana:

All of our sorrow is real, but the atoms of which we are made are indifferent.[2]

It is, to paraphrase the composite French mathematician, Bourbaki, a beautiful problem – one that can be stated simply and will, no doubt, have a very complex solution:

How do we construct real sorrow from hypothetical indifferent atoms?

It is possible that there is no sharp demarcation between the categories conscious and non-conscious (just as we would say today that no sharp demarcation exists between living and non-living). It could even turn out (as shocking as this might appear) that we

---

2 Santayana, G. (1925). *Dialogues in Limbo*, New York: Charles Scribner's Sons.

must invoke a new "law of nature": follow Descartes and pour the conscious substance (the dark matter of neuroscience?) into the machine.

But the conservative scientific position is to attempt to construct this very subtle property from materials available – those given to us by physicists, chemists, and biologists (as has been done many times before: celestial from earthy material, organic from inorganic substances, temperature from the motion of molecules, light from electricity and magnetism, or genes from atoms). The unrepentant reductionist believes that this construction can and will be made and that it will in no way diminish the value or the significance of what has been constructed.

Success would, no doubt, be magnificent, but failure might be more so. If we cannot perform the reduction then we will genuinely have made one of the profoundest discoveries in the history of thought – consequences of which would shape our conception of ourselves in the deepest way.

Richard Feynman, a while back, commenting on the problem of superconductivity, said (I paraphrase): "We don't need any more experiments; that's like looking at the back of the book for the answer. We need more imagination." For the problem of consciousness, we could, perhaps, say: "We don't need any more experiments; we need the leap of imagination that will show us how to construct feeling entities from non-feeling atoms – conscious systems from unconscious neurons."

With such a construction, the rest, while possibly complicated, is conceptually straightforward; without it, the rest is façade. And if the construction is not possible, the consequences would be monumental.

On the Nature and Limits of Science

PART THREE

On the Nature and Limits
of Science

19

What Is a Good Theory?

Is a correct theory necessarily a good theory? The strength of theoretical structures lies not so much in their correctness, but in how concrete they are as well as the precision with which questions can be formulated.

This essay is based on an excerpt from the paper "The BCM Theory of Synapse Modification at 30: Interaction of Theory with Experiment," by Leon N Cooper and Mark F. Bear, originally published in Nature Reviews Neuroscience, in November 2012.

The usefulness of a theoretical structure lies in its concreteness and in the precision with which questions can be formulated. The more precise the questions, the easier it is to compare theoretical consequences with experience. An approach that has been very successful is to find the minimum number of assumptions that imply as logical consequences the qualitative features of the system that we are trying to describe. If we pose the question this way, it means that we agree to simplify. As Albert Einstein once said, "Make things as simple as possible, but no simpler." Of course, there are risks, and it is here that science becomes as much art as logic. One risk is that we simplify too much and in the wrong way, that we leave out something essential. This is the intellectual risk. Another risk is political – that we may choose to ignore in the first approximation some facet that an important individual has spent a lifetime elucidating. (Of course, temporarily setting this facet aside does not reduce its importance. Once we have achieved the initial (zero-order) scaffolding, this facet may provide an essential variation.)

The task, then, is first to limit the domain of our investigation, to introduce a set of assumptions concrete enough to give consequences that can be compared with observation. We must be able to see our way from assumptions to conclusions. The next step is experimental: to assess the validity of the underlying assumptions, if possible, and test their predicted consequences.

What is necessary to make connections between assumptions and conclusions? If one or two steps are all that are required, little mathematics is needed. However, if what is required is a long chain of reasoning with a quantitative dependence on parameters, mathematics, while possibly not required, helps. (It is important to distinguish such mathematical structures from computer simulations, where the preciseness of assumptions is often lost in mountains of printouts.)

A "correct" theory is not necessarily a good theory. It is presumably correct to say that the brain, with all of its complexity, is one consequence of the Schrödinger equation (the basic equation of quantum physics) applied to some very large number of electrons and nuclei. In analyzing a system as complicated as the brain, we must avoid the trap of trying to include everything too soon. Theories involving vast numbers of neurons in all their detail can lead to systems of equations that defy analysis. Their fault is not that what they contain is incorrect, but that they contain too much.

A theory seemingly can be characterized by what it does not predict (and thus can be falsified). But this can be illusory. Additional ad-hoc hypotheses can almost always be introduced to save the original idea. (Think of the use of epicycles to modify circular orbits in the Earth-centered universe). But ad-hoc hypotheses (even though occasionally correct) become increasingly unattractive. Thus we add esthetic criteria: that the theory be beautiful, natural, and elegant with few assumptions, possess a rich structure, and be in detailed agreement with experience.

Most great theories are at best partially correct. The famous Bohr atom was grounded on inconsistent assumptions and was never successfully extended beyond hydrogen. It did, however, give us a stunning, if approximate, account of the spectrum of light emitted and absorbed by hydrogen atoms. One possible position (taken by some) was that because of its deficiencies, it was not worth pursuing. Another (in retrospect, very fruitful) position was that such remarkable agreement between the theoretical structure and what was seen could not be accidental. Thus, the useful question was: What are the consistent new assumptions that yield in the relevant domain, as one consequence, the structure of the Bohr atom?

A theory is not a legal document. In spite of occasional suggestions to the contrary, no scientist is in communication with the Almighty; we possess no tablets delivered from Sinai. Theoretical analysis is an ongoing attempt to create a structure – changing it when necessary that finally arrives at consequences consistent with our experience.

The most important characteristics of a good theory are precision and concreteness. It should be well defined and precise; to paraphrase Galileo, what is said should depend on what was said before. This is not to say that a theory cannot be amended or modified. Indeed, one characteristic of a good theory is that one can move about in the structure – tinker, change assumptions – and know what the new consequences will be.

But in many ways asking what makes a good theory is like asking for a general prescription for a good painting or a good piece of music. We can list many rules (to which there are always exceptions), but in the end you know it when you see or hear it.

20

Shall We Deconstruct Science?

Science and the so-called "scientific method" have been deconstructed and critiqued by many in the humanities. Do we have to redo science or is it okay as it is?

This article based on a talk given at a Partisan Review conference "Breaking Traditions 1896 and 1996," and subsequently published in Partisan Review, 64(2), in 1997.

As a scientist (a hard scientist with a soft heart) before this audience, I believe it is appropriate to quote a remark made by Ed Bloom, a departed colleague. Once, when addressing the Modern Language Association, Ed said he felt like a lion thrown to the Christians.

I was very much influenced by Freud's writings on psychoanalysis, particularly by his theory of dreams. I didn't believe all of Freud's elaborations, but I thought he touched on concepts that were enormously deep, such as the unconscious, the conscious, and the very powerful interaction between them. It is of great interest to me since biologists and neurophysiologists expect to find the underlying physical correlate for such concepts as the unconscious. For example, we might say that the unconscious is related to stored memory and what is conscious is memory in play.

I seem to remember that Freud believed we would eventually find the biological basis for psychoanalytic concepts. But at the beginning of the past century that time had not arrived. So the best way was to proceed without such a basis. He was right. One of the things a scientist learns is that there is a right time to ask a question. Newton did very well in his analysis of the motion of the planets but failed completely in his attempts at chemistry or the nature of matter, because their time had not yet come. My own feeling is that within the next ten to twenty years we will uncover the biological basis of learning and memory storage. We will know precisely what happens in a cell or at the cell surface, which receptors change, what alterations take place, what cellular changes occur with learning, what chemical changes take place, where memories are stored, what the differences are between short-term and long-term memory, and what determines whether or not transfers are made between short- and long-term memory. I believe that it is just a matter of time before we understand how the brain processes information.

A question we might ask is whether we will be able to construct all of the so-called mental states from these material entities. Whether we will be able to construct such things as our awareness of ourselves, our consciousness. I recall Santayana writing (*Dialogues in Limbo*, I believe): "All of our sorrow is real, but the atoms of which we are made are indifferent."[1] The scientific question might perhaps be phrased: Can we construct real sorrow from indifferent atoms? This is no doubt difficult, but not necessarily insoluble.

I'm not willing to indulge in the kinds of mystification one so often hears – the almost invariable accompaniment to the players who attempt difficult scientific questions. There is a common sequence. First try to solve the problem. After the first few failures throw up your hands and say it is insoluble or that a new law of nature is involved. In this case pour intelligence and consciousness into the material machine. We don't know before a problem is solved whether or not it is soluble, but historically people have thrown up their hands very quickly.

---

1 Santayana, G. (1925). *Dialogues in Limbo*, New York: Charles Scribner's Sons.

Let me give you an example. In the nineteenth century, the French positivist Auguste Comte said that an example of knowledge permanently inaccessible to the human mind is the chemical composition of the stars. He likely reasoned that in order to determine the chemistry of the stars you have to go there, take a sample of a star and put it in a test tube, etc. That was just a couple of years before spectroscopy was invented. Then it was realized that by analyzing the light that came from the stars you could know more about the chemistry of the stars than we do of the interior of the Earth. What it really means when people say that things are insoluble is that they haven't seen the links: the links don't yet exist. But links are what we construct all of the time. It could turn out that some problems are insoluble, but I'm reluctant to jump to that conclusion.

At the turn of every century we seem to be afflicted with *fin de siècle* malaise; call it "sixth decimal place-itis." You recall the famous statement made at the end of the nineteenth century that all that's left in physics is filling in the sixth decimal place – just a few years before we had relativity, quantum theory, and so on. At the end of the twentieth century there were authors writing that history had come to an end, that science had come to an end. All that is left is ironic science, whatever that is; that the kind of questions we deal with today can never be experimentally verified. From the point of view of a working scientist nothing could be more absurd. We all think that the questions we are working on can be verified, that we will verify them and will solve the problems just as our predecessors solved theirs. In fact, I remember when I was a mere youth at the Institute for Advanced Study, all the young guys would sit around saying Heisenberg, Schrödinger, Einstein – they solved all the easy problems and left us with the hard ones. In the last twenty or thirty years most of the problems that we considered hard have been solved, in many cases by us. I'm sure young people sit around now saying we solved all the easy problems, leaving them the hard ones. Of course, once you know the solution, the problem looks a lot easier than it did before.

I tend to be optimistic. I think science plugs along, does its thing, and gets results. Somehow, because scientists like me perhaps are down-to-earth, cheerful, and optimistic, and believe, deep down, that there really are no questions we can't answer, we have become the subject of intense attacks. Perhaps it is science envy. Perhaps it's indigestion. But I find some of the words I read hard to believe. As though they came from another planet. Still, I find that people rarely say things face to face that are quite as shocking as what they print. And often, there is a point to the criticism. Turning Voltaire upside down, I tend to sometimes agree with what is said, and disagree with the way it is being said. Let me address some of these concerns about science.

You've all heard of the famous "scientific method," maybe have even written about it. I've done science all of my life and don't have a totally clear idea what this method is. A real scientist, I would say, might be compared to Sherlock Holmes: he has a good nose for where to go, and sufficient technique to get him there. I've known scientists with remarkable technique who always seem to be pointed in the wrong direction, others with minimal technique who always seem to be going the right way. We come in all varieties. Think of Sherlock Holmes again. He refuses to speculate because he doesn't have enough data; but when he has the data he produces the miracle – he ties everything together. To characterize science in a few words, one could say that the job of scientists is to get the facts (as problematic as they are) to distinguish the world that is, from all the ways it might be – from the world of our fantasies, from all other logical possibilities, to determine which world we are actually living in. Then (and of course we are aware that facts interact with theories), introduce a theory, a structure that ties everything together.

The theoretical structures we have produced (allow me to wax romantic) are the intellectual cathedrals of the twentieth century, arches from the origin of the universe to the properties of matter, from coldest to hottest, calculations that yield the amount of magnetism on an electron to ten decimal places. It is a terrible pity that so few people understand and appreciate what is there. The idea that it is too hard (all that mathematics) is absurd. It is not any harder than learning a second language. We must distinguish between the talent and virtuosity required to create science, and the ability to be led through it, to appreciate what is there. We can watch a tennis match and understand what is going on, but we can't get on the court with Pete Sampras. It does take special talent to create science, it's true, but it's not difficult to understand it. That holds true for all the science I know, including the most esoteric aspects of the quantum theory.

We may also have heard (I'm shocked) that scientists don't always follow the scientific method. Scientists are influenced by a current point of view, so that what they are doing is relative, therefore no better than other ways of knowing. I classify this, in my own unscholarly way, as a school of thought that says everything is as good as everything else – total absurdity. Because something is not the best or the worst does not mean that a distinction cannot be made. Life consists of choosing between things that are a little bit better and a little worse; that there are no distinctions because there is no absolute truth and no absolute falsehood is like saying that the most magnificent lines written by Shakespeare are equivalent to anything that comes out of anyone's mouth.

But I do agree that there is a point of view in science. It is important to teach science in a historical context, because the questions people ask are not really comprehensible unless you understand what was believed at the time. Let me give you an example. You may have heard of the Michelson–Morley experiment. This was done toward the end of the nineteenth century by the famous American physicists Albert Michelson and Edward Morley. The results of this experiment were startling to physicists and finally led to a complete revolution in our concepts of space and time. (It led to Einstein's special theory of relativity although, as one of history's ironies, it might be that Einstein was not aware of this experiment in 1905.) What was this experiment? They thought of the Sun and

Earth moving through space (the luminiferous ether?) and asked what was the motion of the Earth with respect to this ether. Michelson's genius was, in part, to have developed the techniques required to measure such a possible motion of the Earth. When he and Morley made the measurements they found, to put it most simply, that the Earth was standing still, that the Earth's motion with respect to the ether was zero. They were dissatisfied because they wanted to get a number.

But the result was unambiguous. Even though the Earth is moving in its orbit about the Sun at about 18 miles a second no matter when during the year they measured, the result they found over and over was that the Earth seemed to be standing still. Many proposals were made to explain this result. But as far as I know, no one ever proposed that the Earth was, in fact, standing still (at the center of the universe?).

Now imagine that the Michelson–Morley experiment had been done before Copernicus. Everyone would have said, "Why did you waste your time? We all know that the Earth stands still at the center of the universe." But at the end of the nineteenth century that was the one thing no one was willing to say. Why not? Because the whole context had changed. Scientists shared a different point of view.

Of course scientists are influenced by what other people believe and by their times. The underlying meaning of a scientific theory can change quite radically from one century to another. When people say that science is not objective but subjective, they do not appreciate that very few scientists worry about underlying meaning in the first place, and when they do so, they generally don't do it very well. This is not their daily concern. Science is a craft in which internal structures are built – as the structure of a cathedral. It is astonishing that between the nineteenth and twentieth centuries that cathedral has been picked up and moved from one place to another. It sits on other foundations, there is another view of nature; but the internal structural relations are essentially the same.

And it is these internal structural relations that are the real content of science, and they remain essentially invariant from Newton to Einstein to the quantum theory and no doubt will remain the same in theories that emerge in the future.

Let's talk a bit now about Alan Sokal, who submitted a paper to *Social Text* called "Transgressing the Boundaries: Transformative Hermeneutics of Quantum Gravity." If the title doesn't make you laugh, what follows surely will. After the editors of this journal published it, they claimed they were deceived. Stanley Fish wrote that Sokal was undermining the trust of one person in another. Perhaps he did to a certain extent. But what is going on if the editors of a magazine can't tell the difference between something that is patently a hoax and something that is real? If someone sends me a scientific paper in a field with which I am familiar, even if I don't understand everything that is written I would hope to be able to determine whether or not it was a hoax. If nothing follows from anything else, I would send it back and ask for clarification. In an electronic posting on the internet Bruce Robbins and Andrew Ross, the editors, were furious, and stated:

Why does science matter so much to us? Because its power is a civil religion, has a social political authority that affects our daily lives, and the powerless condition of the natural world more than does any other domain of knowledge. Does it follow that non-scientists should have some say in the decision making prophecies that define and shape the work of the professional scientific community? Should non-experts have anything to say about scientific methodology and epistemology? After centuries of scientific racism, scientific sexism, and scientific domination of nature one might have thought that this was a pertinent question to ask.

I freely admit that science does have a point of view, but not that our point of view is racist or sexist, or dominates nature – it dominates our thinking about nature, but nature is what dominates our structures. However, there is an issue: science is powerful and affects our lives; an important question is whether non-scientists should have anything to say about science. That is an interesting and legitimate question to which I would like to propose an answer. When it comes to the methodology, when it comes to the way science is conducted, clearly it is scientists, the experts who have to decide. To say that non-experts should decide is like saying that a passenger on a plane should be asked for his opinion on what the pilot should do when in a difficult situation. Pilots may make wrong decisions, but even if they don't always do it right, they have the best chance.

But where values are involved, when the issue is the way science affects society, there are situations that should not be decided by the experts. They must involve everyone. Here we get into really serious problems. People feel so frightened and left out because they often are excluded from the decisions that affect their own lives by experts who say "You just don't know enough." I think that is too bad. To know enough to be able to assess the risks and to make an intelligent decision about many of the questions that are asked is not as difficult as is believed.

Consider the recent brouhaha about electric lines and spending billions of dollars to move them, etc. Some of the discussion reflects what can best be described as bleak ignorance. It really doesn't take more than a week to learn enough electromagnetic theory to comprehend why those of us who understand it believe that the supposed adverse effects are not there. Of course anything is possible. But do we want to protect against hypothetical possibilities for which we have no evidence whatsoever? We have to assess risks. And it's not all that hard to obtain knowledge so that we can all participate in decisions. I think non-experts must be involved in these decisions. I think it is possible for non-scientists to learn what they need to know about science. Science departments (with some outstanding exceptions) have probably not been as responsive as they should. But it can, it must be done. In this world ignorance of science is as much a handicap as illiteracy. What is required is not "science studies." What we need, instead, is that all of us study more science.

The following are my responses to questions that I believe will be evident from the responses.

Do facts exist? An old and deep philosophical question, "Does the external world exist?" gets us right back to Bishop Berkeley, even Plato. My own view is that, of course, our perceptions of the world are influenced by our own state of mind. One of my colleagues once said, "People say that seeing is believing, but in fact in many cases you have to believe in order to see." That's perfectly true, but to paraphrase Orwell, all facts are equal but some are more equal than others. For example, if you shine sunlight through a spectroscope and look at it carefully you will see certain dark lines. These are the famous Fraunhofer lines and, as far as I know, almost all people who look will agree that the lines are there. They will be there in specific places whether the observers are sad or not sad, etc. I think we can agree that events do exist that most people more or less agree on, even though they may influence them in very different ways. When we construct a theory of knowledge the question might be: "Assuming that there is an external world, how can we organize it with a mind that's so imperfect?" I know there is a problem, but I think generally we can distinguish "facts" from our feelings about them.

Another issue. You stated that there might be physiological bases for various mental states, the various concepts of analysis, but you weren't sure if there would be any use to get them; I have the feeling that you don't even like the idea. But whether we like it or not, we'll very likely find that what we call "soul" is some configuration of material objects. I'm not saying I particularly like it or dislike it, but we might as well face the fact that it is probably that way.

One thing you said that I agree with completely: one will not have a theory and understanding of mental states until one can construct mental states from material objects in such a way that the construction displays all the properties of the mental state. We have done it before: we have constructed such things as temperature from the motion of molecules; we have constructed light from electricity and magnetism. That's the business we're in; if we can do it, we will have to satisfy the criteria that you set. We don't know yet whether we can do it – let us just see.

Scientists are human beings. When I was a teenager, my mother read an article in the *New York Times* in which some scientist was caught in an embarrassing position; she told me it shocked her. Teenage wise-guy that I was, I said, "So what, everybody does it," and she said, "But he is a scientist."

It doesn't seem especially surprising to me that scientists lie, steal, and cheat the way everybody else does. And before you all, I declare myself against lying, stealing, and cheating.

But what should we do if a scientist falsifies data? I can assure you none of us in the game like that, because it can waste time and drive us crazy. The question is how to deal with it. The worst thing to do is to bring in lawyers and the FBI. The best thing to do is just to repeat the experiment, that's all. Just do it over again. Fraud is not the problem. We have to decide in every article we read whether the experimenter is reporting exactly what he saw or what he thought he saw, or what he believed he ought to have seen. We worry about that all the time. Of course, if someone deliberately falsifies, that can be aggravating. But it happens rarely and it is self-correcting. If a scientist reports experiments that no one else can repeat, very soon his reputation sinks below that of fraud – to incompetence.

No matter what ideas or themes we propose, there is always someone who says you could find something similar in a footnote here or there. There are always predecessors, ideas already in the air. If you were to take an extreme point of view, nothing new could ever happen. Certainly the world is different since Freud, so he must have done something. It is what scientists do, putting things together that make a difference in the world.

Scientists are not value free. Many of us would hesitate to work on harmful or destructive applications. But the context is important. In a wartime situation (World War II for example) when it was feared that the Germans might be ahead of us in the construction of a nuclear bomb, we worked frantically to deny Hitler the possibility of threatening us with this new weapon before we had it ourselves. When it comes to fundamental research, we scientists are no better than others at foreseeing the consequences of what we do. We don't have a crystal ball.

We are occasionally presented with a romantic image of the scientist (Durrenmat or Hollywood) on the verge of the momentous discovery that can lead to limitless evil who says no – I will not go on. Perhaps Oppenheimer encouraged that image. But mostly scientists grope along with no clear indication of what the consequences will be. I see no directive within science that can guide us. We, as citizens and humans, must decide.

I agree with you that we would be better off sharply limiting the number of nuclear weapons right now, but nuclear weapons were created under the pressure of World War II and were made possible by scientific advances that date all the way back to the beginning of the century when most scientists wouldn't have dreamed that nuclear weapons would be possible. In fact, Rutherford once said the idea that you can get energy from the nucleus is moonshine, and he was one of the greatest nuclear physicists who ever lived.

This is a bit like asking whether a detailed, chemical knowledge of what happens when we share a bottle of Lafite Rothschild is going to change our appreciation of that great wine. One has nothing to do with the other. I would love to have a thorough understanding of everything, but I don't expect that to change my appreciation of my own experience. I don't believe that anything we learn about ourselves will relieve us of personal responsibility and accountability. Of course, you could cite extremes in which a person cannot function or make choices. But for most of us choices are possible. I don't believe anything we learn will change that.

21

Visible and Invisible in Physical Theory

How is it that physicists can invent equations, so rich in structure, so detailed in consequences, and so closely in agreement with experience? What does it mean when we create equations that agree with experiment to ten decimal places? Do these equations have a reality of their own?

This essay is based on a lecture originally presented at the Trieste conference in honor of Paul Dirac in 1972.

It is more than fifty years since Paul Dirac gave us one of the most beautiful inventions of the twentieth century:

$$H = c\mathbf{a} \cdot \mathbf{p} + \beta mc^2 \tag{21.1}$$

I once asked Professor Dirac how he came upon his masterpiece. If my memory serves (I hope he will correct me if it does not), he said that in the late 1920s most physicists, trying to calculate with relativistic quantum mechanics, were content to use the Klein–Gordon equation modified with ad-hoc corrections. In his attempt, Dirac did not try to put in the electron spin but rather to treat in a consistent manner the simplest possible problem: that of a spinless, relativistic electron. The critical moment came when he realized that 4 × 4 matrices were required. "How long did it take you from that point?" "About two weeks." I couldn't refrain: "why so long?" He was a bit taken aback; then, his attention focused perhaps on those two glorious past weeks, he nodded as though to agree in silence – why so long?

Everything is quick today. High-school students solve the Schrödinger equation. College freshmen manipulate γ matrices and undergraduates draw Feynman diagrams. Functional integrals are reserved for first-year graduate students. The facility is so great, the technique so smooth, we can easily forget one of the most remarkable facts about Dirac's great equation: its persistence in spite of vastly changed interpretation – from single-particle relativistic Schrödinger equation to the equation for a relativistic spin-1/2 quantum field.

This leads to a broader – perhaps unanswerable – question: What is the significance of those mysterious equations the happy ones among us invent – those not immediately visible rules that reside behind the visible façade of "real" experience? Do they have a reality too? It has astonished many that the theoretical structures we create should correspond so closely and in such detail to what nature is. Albert Einstein perhaps expressed the common wonder with his remark to the effect that what is most incomprehensible about nature is that she is comprehensible.

A possible view is that what we might call "laws of nature" have a real existence – as though they were written down in a manual of the universe to which, of course, we have no direct access. If this were the case, it might be reasonable to suppose that, perhaps by working very hard, we would one day learn what they are. That, I think, is difficult to believe literally; there surely is order in nature. (Or there surely was order: the Sun may not rise tomorrow, but it has risen fairly regularly most mornings in the past.) However, is the detailed nature of this order something we discover? Or is it something we invent?

It is my opinion that a large part of what we call physics is invented. We see this most clearly in those profound conceptual changes – for example, from classical to quantum physics – in which the fundamental axioms and meaning change (the deterministic classical theory replaced by the non-deterministic quantum theory) yet the structure of the two theories in some domains (e.g. the planetary orbits) is almost identical. (If for some reason we could never experience "quantum" phenomena, which theory would be "true"?)

There is a sense in which the scientist is imposing an explicit ordering created in his mind on a more or less recalcitrant nature. In the perspective of today's quantum world, consider the mechanistic Newtonian ordering, as seen, for example, by Laplace:

If an intelligence, for one given instant, recognizes all the forces which animate Nature, and the respective positions of the things which compose it, and if that intelligence is also sufficiently vast to subject these data to analysis, it will comprehend in one formula the movements of the largest bodies of the universe as well as those of the minutest atom: nothing will be uncertain to it, and the future as well as the past will be present to its vision. The human mind offers in the perfection which it has been able to give to astronomy, a modest example of such an intelligence.[1]

Returning to our question: Is there a sense in which there is something beyond the visible? If so, what is that sense? Are causes linked to effects by relations that really exist in nature as well as in our heads? I do not think I will answer these very old questions in this essay, but in the context well known to physicists, they can be put very precisely: How is it that we can write equations so complex in form, so rich in structure, so detailed in consequences, and so closely in agreement with experience?

Consider something we all know very well: Maxwell's equations of electricity and magnetism. They have a very rich structure and consequences that are in agreement with classical electromagnetic phenomena. With these equations, Maxwell identified light as an electromagnetic wave, calculated the speed of this wave using known magnetic and electric constants, and found that it was in agreement with the measured speed of light. It seems somewhat of

1 Laplace, P. S. (1840). *Essai Philosophique sur les Probabilités*, Paris: Bachelier, p. 4.

a miracle that a set of equations so complex, so detailed, could be written down and describe nature so accurately.

This miracle is compounded when we replace Maxwell's classical theory by our present best guess: quantum electrodynamics, a theory whose assumptions are remarkably different from those of Maxwell but which, with some few exceptions, gives an essentially identical structure in the realm of classical phenomena. What is it that makes it possible for them both to be in such close agreement with each other and with what is actually observed?

From one perspective, this may not seem to be particularly mysterious; quantum electrodynamics can be regarded as a generalization of Maxwell's electrodynamics so arranged that, by a correspondence principle, Maxwell's theory is retrieved in the classical domain. But this is primarily a historical statement. We might imagine that on two different planets (or in two different ecological niches), one biased toward quantum phenomena, the other toward classical, these two theories could be independently invented.

A modern point of view concerning physical theories suggests the beginning of a possible answer. Since Einstein's 1905 paper "On the Electrodynamics of Moving Bodies," physics has become almost completely dominated by a search for symmetry principles. Einstein's 1905 paper contained a new and major directive to theoretical physics: All theories must obey the principle of relativity (be Lorentz invariant). To this we have added various other symmetry principles: gauge invariance, symmetry under interchange of identical particles (Fermi or Bose statistics), SU (...), etc.

Consider the symmetries of Lorentz and gauge invariance. If we ask for a quantum theory of a fermion that obeys these two principles, we almost immediately write down quantum electrodynamics. It is true that we can also write more complicated theories (e.g. for particles of higher spin), but we can agree that the Dirac electron interacting with a massless vector field, quantum electrodynamics is among the immediate and simplest of the possibilities.

Classical theories are not as constrained, but again, given a particle and a massless vector field that interact in such a way as to obey the principles of Lorentz and gauge invariance, we obtain almost immediately classical electrodynamics.

We might say then that if we knew these two principles and we knew further whether the world were classical or quantum, we could almost immediately write down either classical or quantum electrodynamics. Not only would these be the actual "correct" theories in the two domains, but they would give the usual almost identical results in the common domain.

One might reasonably ask: "Is this really simpler than having discovered Coulomb's law, Ampère's law, Faraday's law, and so on??" (We are reminded that it took Maxwell close to a lifetime to arrive at his masterpiece.) I am not sure it is simpler and, until very recent times, the path of discovery has led from specific "laws" to the symmetries they almost accidently seem to obey.

However, it may be that these "underlying" symmetries are what there is in nature to be discovered. We might say that it is the principles of Lorentz and gauge-invariance that unite classical and quantum electrodynamics. One is the simplest classical theory, the other the simplest quantum theory that obeys these two principles. Can we imagine a Lorentz and gauge-invariant xxx theory, where "xxx" is what may eventually replace "quantum?"

It may be that in our local sampling ("local" because we can experience only a very limited region of space and time) what we can discover is a kind of local symmetry or geometry. Once obtained, we construct the simplest objects consistent with these restrictions on our region of space-time. We recall that the simplest objects we can construct are not necessarily all found in nature: free quarks have not yet been seen; a massless vector field is present in electrodynamics, but is apparently not among the strong interactions. Now, such facts may be profound consequences of the proper symmetry principles in a theory with a superposition principle (that can lead to broken symmetries and related phenomena), but we still can only vaguely guess how.[2]

However, it is strikingly true that all of the "elementary" objects we find in nature (in fact, some that have not yet been found) – those objects that are the real subject of much of physics – are among the simplest that are consistent with that we believe are the local symmetry properties. All of the complexity of the world – objects that are consistent with, but in no way simple realizations of the underlying symmetry, and in recent work may not even manifest the symmetry – is imagined to be constructed of various combinations of the elementary objects. If we were asked: What is the physics of a 100 000 000 share day on the New York Stock Exchange, one would probably respond that somehow that is not our concern; it is too complicated. In physics we have always attempted to extract what we call fundamental objects. This selection may be just the point. What we are really saying is that it is possible to find certain kind of objects consistent with the underlying symmetries and that everything else found in nature is constructed of them.

Let me illustrate with a simple example. Suppose we were trying to discover the geometry of our space: not the dynamics, just the spatial relations. It is clearly possible to construct areas or volumes of almost arbitrary complexity. There will exist relations among them (even in this very simple situation) of bewildering complexity. Some of these might or might not exist in the "real" world. Now the liberating idea would seem to be that locally our space is in some sense rotationally and translationally invariant and that a Euclidean metric is at least approximately appropriate. If objects in the real world can be found that behave more or less like lines and points, we can construct triangles, etc. and find various and possibly very intricate relations among them. We are then on our

2 Recent theoretical advances have clarified this situation.

way to creating a logical structure in agreement with what is in the world.

Let me note further that the symmetries we propose are proposed to be exact, not for the actual but for an idealized world. (The actual universe is clearly not rotationally or translation invariant; it is the Lagrangian that is thought to be invariant, not necessarily the state vector.)

Think of this in the context of trying to discover the geometry of the surface of the Earth. For a variety of reasons known very well to us, the Earth is not exactly a sphere. And most of it is not very smooth. Yet any geometrical theories consistent with the symmetry and metric of the surface of a sphere would give a remarkably good description of a very large variety of phenomena concerning spatial relations over large portions of the surface of the Earth if one chose to look at the appropriate phenomena (the oceans on a large scale for example). In addition, one could write intricate equations that gave detailed and accurate relations if one chose the proper phenomena to analyze.

Let me conclude with a final comment: although symmetries such as Lorentz and gauge invariance have great simplicity, we know they are not evident, nor were they easy to discover. However, there is a sense (think of rotation or translation invariance) in which symmetries are more visible than the complex theories that satisfy these symmetries. It may be that we can extract from nature approximate symmetries (perhaps locally very close to exact) and that physical theories that satisfy these symmetries are capable of very accurate descriptions of large portion of nature and will be in agreement with each other in certain domains because they satisfy these symmetry principles. It could turn out that what we think is deepest is closest to being visible.

22

Experience and Order

Where does order come from? Is it there in nature to be discovered, or do we invent it? Do we discover the laws of nature or do we create them? Is there a difference between these two viewpoints?

This essay is based on a chapter originally published in An Introduction to the Structure and Meaning of Physics *in 1968.*

Man comes into the world with a cry: a burst of light, a slap, initiate him into the universe of sensation. The material of science is this: our experience of the natural world, the world that is – not the ones that might be. Somehow in the mind this raw experience is ordered, and this order is the substance of science. What happens there is the application to a great diversity of the phenomena of the world of many of those elements called common sense, used every day and resting on certain suppositions we make concerning the world about us. Some of these are probably universal to man and beast. Others are more particular. We tend to accept them without special awareness, and some are so well hidden we are scarcely conscious of their existence.

The foremost supposition is the belief that the world outside ourselves, outside our own mind, exists. This belief is so primitive that it is very likely shared by all, except animals lowest on the evolutionary scale and some philosophers (whose position on the evolutionary scale we cannot guess). It may be that a newly born child is not aware that the patterns of light, sound, touch, smell, and taste to which he is exposed have their origin in objects outside his mind. He may not know where he ends and something else begins. The first realization that an often-repeated pattern of sensation is another person – mother – is then a discovery whose magnitude is never equaled. Yet, it is a discovery all of us who grow up and function make.

What is called a chair is a concept created to unify what we see from one side, and then from another; what we feel with our hand, and what supports us when we sit on it. We believe without analysis that all these sensations proceed from a single object, the chair. There is no additional evidence for its existence. We might experience the same sensations in a dream or hallucination in the absence of an actual chair. When we propose the existence of any object (electron, chair, or neutrino) it is done to unify such a variety of experience. This is probably the most primitive, yet the most important, theory we create.

The identification of a single object as the agent for various sensations is not always easy. For a long time men thought of the morning star and the evening star as two separate celestial objects. It was probably an astronomer of Babylon who first identified them as one and the same – the planet Venus. Western mountaineers report the surprise of their Sherpa guides when they are told that peaks they have seen in different profiles for their entire lives are many faces of the same mountain.

As we climbed into the valley we saw at its head the line of the main watershed. I recognized immediately the peaks and saddles so familiar to us from the Rongbuk (the north) side: Pumori, Lingtren, the Lho La, the North Peak and the west shoulder of Everest. It is curious that Angtarkay, who knew these features as well as I did from the other side and had spent many years of his boyhood grazing yaks in this valley, had never recognized them as the same; nor did he do so now until I pointed them out to him.[1]

---

1 Shipton, E. (1965), quoted in J. Brownowski, *Science and Human Values*, New York: Harper & Row, p. 30.

An infant rapidly develops expectations about the sensations to which it is exposed (to the sorrow of the over-indulgent parent). A child usually does not have to be burned twice to learn to keep his fingers away from fire. Without analysis, he believes that there is order in the world; it is not difficult to understand why. In a world where similar things repeat, the animal that is so made that it can grasp this fact survives more easily than its fellow who cannot or will not. The animal philosopher, whose mind is sophisticated enough to argue – "The tiger ate my brother but that does not permit me to conclude that he will (given the chance) eat me" – increases the likelihood that he will provide a meal for the next hungry tiger. One's view of the world thus affects one's chances of survival. Those who can adapt themselves to the order (just as do those who can adapt themselves to the temperature, moisture, and other environmental conditions) in the world survive best.

We all share this lct us call it instinctive – beliet in the order of the world. But our response to the belief is quite varied, and it is in this response that men differ among themselves, and in which animals differ so much from men. Thinking – according to Pierce, an activity that begins when the mind is uncomfortable and ceases when it is comfortable again – has led man in many diverse directions, sometimes enlightening, other times not. To the animal mind – at least as one might imagine it – simple associations occur: a paw placed on a hot coal is burned; thereafter, hot coals are avoided. A bell rings and food comes; and salivation after a time begins when the bell rings. Whether animals venture further we do not know, but without prejudice to our fellow creatures we may imagine a state of mind in which no questions are ever asked.

Men, however, being featherless bipeds, seem characteristically unwilling to remain in this passive condition. Demons, spirits, purpose, or the machine-like laws of nature have been proposed to "explain" the association of two events. Once explanations begin, frequently what has actually been observed becomes confused with the explanation. A belief, strongly implanted, has a life of its own.

Man, alas, is mortal, but ideas and superstitions seem to live for ever; and when they become involved with our emotions, it takes a rare ability to distinguish what we see from what we believe should have been seen.

It seems simple to see what the world is, yet it requires the perception that hopefully comes with age, together with an innocence sentimentally attributed to children. A child, for example, pushes a switch on a television set and sees moving images appear on the screen. To him, this is as plausible as anything else; he is surprised only when the switch is pushed and the screen does not light up. To the young mind, any relation or correlation is as believable as any other, because it has no predisposition. It is no more remarkable that a button should be pushed and a picture appear than that a mouth should open and a voice be heard. If there is such a thing as the much-mentioned scientific method, one element certainly would be just this openness and honesty – the interest, the skill, and the commitment to look at the world as it is, a commitment expressed by Descartes when he said that his library was the calf he was dissecting, by Galileo when he looked at the heavens through his telescope, or by Aristotle when he watched the bees:

There is a kind of humble-bee that builds a cone-shaped nest of clay against a stone or in some similar situation, besmearing the clay with something like spittle. And this nest or hive is exceedingly thick and hard; in point of fact, one can hardly break it open with a spike. Here the insects lay their eggs, and white grubs are produced wrapped in a black membrane. Apart from the membrane there is found some wax in the honeycomb; and this wax is much sallower in hue than the wax in the honeycomb of the bee.[2]

It would include, further, a commitment to compare one's theoretical conceptions with what is in fact observed. For, as Aristotle said more generally, in a critique perhaps intended for one of his better-known instructors:

2 Aristotle, *Historia Animalium*, Book V, Chapter 24.

Lack of experience diminishes our power of taking a comprehensive view of the admitted facts. Hence, those who dwell in intimate association with nature and its phenomena grow more and more able to formulate, as the foundation of their theories, principles such as to admit of a wide and coherent development; while those whom devotion to abstract discussions has rendered unobservant of the facts are too ready to dogmatize on the basis of a few observations.[3]

However, seeing is not easy when belief is strong (and we live in a world with no lack of strong belief). The so-called man of science must, therefore, develop an objectivity that can separate him from his contemporaries and, like the artist, make him seem an observer. Jenner's discovery of a vaccine against smallpox, tearing from a morass of opinion and superstition the fact that the girls who milked the cows did not catch smallpox, required an eye like a painter's, an eye able to see what is, and to separate that morsel from all that is believed. It is said that when Thales told the natives of Miletus that the sun and stars were fire, they looked at him in astonishment, for they worshipped them as gods. And when Galileo said that the sun had dark spots and that Jupiter had satellites, some of his contemporaries would not even look; he wrote to Kepler: "Shall we laugh or shall we cry?" In this aspect, the face of science is hard. For it must, if it is to be at all, look at the world without illusion, just as in the greatest works of art the world of human emotion and experience is viewed as it is, without sentimentality.

This need for trustworthy observation, the raw material from which we construct our view of the world, has driven the scientist into the laboratory and has created the public image of the man in the white coat. The laboratory is a place where certain kinds of observations can be carried out more accurately and perhaps with less distraction than on a city street. The readings on a meter may seem impersonal and objective, but they are so only if the eye that watches and the mind that interprets are themselves objective. For

3 Aristotle, *De Generatione e Corruptione*, Book I, Chapter 2.

the man with a theory, a career, a reputation at stake, the flickering of a needle on a meter generates enough emotion so that objectivity has occasionally been strained.

If a man attempts to measure the temperature of water with his hand, he may or may not estimate it accurately. If he measures the temperature with an instrument, he may gauge it more precisely. In either case he might be right or wrong; but we have learned as a matter of experience that an instrument, properly built and maintained, will give a measurement of temperature that is usually more accurate than that of a hand placed in water. We can measure time roughly by our pulse beat, as Galileo once did. But there is a limit to the accuracy of such a measurement, a limit than can be improved by using a pendulum, a watch, or a very modern clock stabilized by hydrogen-atom vibrations, losing only about 1 second in 100 000 years.[4]

The necessity for measurement pervades science for several reasons. One is the recognized need to separate our observations from our beliefs; the use of machines, impersonal machines, is one way to do this. But we must always remember that a skillful eye often can achieve this separation as well as any machine. A second reason is the obvious desire for more accurate measurements than our senses can produce unaided. And a third reason is the usefulness of measurements that can be duplicated in different places and by different people; machines that can be exactly reproduced facilitate this duplication. Thus, when we say that science weighs and measures, what we mean is that science attempts to achieve a knowledge of the world as it is, a knowledge that is at once accurate and public, so that men all over the world can duplicate it if they create the proper circumstances.

But the gathering of facts without organization would yield a filing cabinet in disarray, a random dictionary, that dull and useless

---

4 The state of the art in atomic clocks has advanced considerably since this was first written. The National Institute of Standards and Technology has since built a clock that neither gains nor loses a second in 3.7 billion years.

catalogue sometimes confused with science. Yet what is there in experience itself to indicate that order can be found? What is it that produces a conviction strong enough to have kept Kepler working for years calculating and recalculating orbits, or Galileo working for a lifetime to understand the motion of bodies? We have no guarantee – in fact, it is a little surprising – that we can find any relations such as those between the orbit of the moon and the path of a projectile near the surface of the Earth. What is there that leads us to believe that an order we might create would be any less complex than the events themselves, that the symbols we write down on paper will somehow permit us not only to know but also to manipulate the world?

There is a mystery in writing, and in times when writing was not common, the act itself was considered magic. Wotan, to learn the mystery of the runes (writing), suffered heroic agonies, as does many a student today. Ode, among the Greeks, originally meant a magic spell, as did the English rune and the German *Lied*. A comic-strip superman tames his enemies by uttering the right formula; another unlocks his powers by saying "SHAZAM." In some of the most primitive magic we can find that effort to attribute mystical qualities to numbers, the conviction that relations between symbols are like relations between objects in the real world, and that somehow the manipulation of the symbols gives us a power over the stubborn material of the world.

From Egypt, in a passage of the *Book of the Dead*, based on a spell of the "Pyramid Texts," comes the "Spell for Obtaining a Ferry-Boat." Professor Neugebauer writes:

The deceased king tries to convince the Ferryman to let him cross a canal of the nether world over to the Eastern side. But the Ferryman objects with the words:

'This august god (on the other side) will say, "Did you bring me a man who cannot number his fingers?" However, the deceased king is a great 'magician' and is able to recite a rhyme which numbers his ten fingers and thus satisfies the requirements of the Ferryman.

It seems obvious to me that we are reaching back into a level of civilization where counting on the fingers was considered a difficult bit of knowledge of magical significance, similar to being able to know and to write the name of a god. This relation between numbers (and number words) and magic remained alive throughout the ages and is visible in Pythagorean and Platonic philosophy, the Kabbala, and various other forms of religious mysticism.[5]

And in Pythagoras, schooled in the mysteries of the Orient, we find possibly the earliest suggestion of a theme that has become the leitmotif of physics: that the order of the world is somehow to be found in orders and relations among numbers.

In the sixth century BC, when the Pythagorean school flourished, numbers as we know them were not in use. Instead, numbers, or relations between numbers, were discovered by placing pebbles or making dots in the sand. These pebbles or dots could be grouped in various arrangements. Two of the arrangements studied were triangles or squares; it was apparently in this fashion that the notion of triangular and square numbers arose and that the relations among such numbers were deduced. For example, an inspection of two square arrays 4×4 and 3×3 leads eventually to the relation

$$4 \times 4 - 3 \times 3 = 2 \times 4 - 1 \qquad (22.1)$$

and finally to the abstract relation (probably written down later)

$$n \times n - (n - 1) \times (n - 1) = 2n - 1. \qquad (22.2)$$

A particular configuration of pebbles that seemed singularly perfect, the monad, fascinated the Pythagoreans, and they made much of it (to the amusement of Aristotle, among others, who came later). This array of pebbles that add up to ten, having survived

5 Neugebauer, O. (1957). *The Exact Sciences in Antiquity*, 2nd edn., Providence: Brown University Press.

Aristotle's sarcasm, reappeared in a letter by thirty-three authors called "Observation of a Hyperon with Strangeness Minus Three" (Figure 22.1).[6]

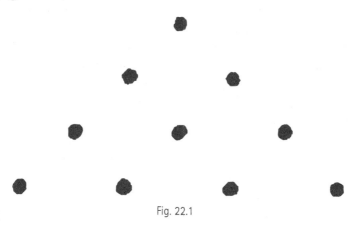

Fig. 22.1

Among the discoveries of Pythagoras was the relation between the hypotenuse and sides of a right triangle, which today we write in the form:

$$a^2 + b^2 = c^2. \tag{22.3}$$

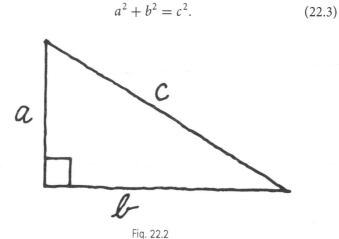

Fig. 22.2

6 Barnes, V. E. *et al.* (1964). Observation of a Hyperon with Strangeness Minus Three, *Physical Review Letters*, **12**(8), 204.

We are told that he sacrificed one hundred oxen to the muses in thanks for this, the most famous theorem of antiquity.

Once, it is said, when he passed a blacksmith's shop, he heard different sounds made by the blacksmith's hammer striking metal rods of varying lengths. He then experienced an astonishing revelation: uniform rods whose lengths are in simple ratios produce harmonic sounds. If one can say that physics began at all, it was at a moment such as that. For Pythagoras realized that the relation between two properties of the real world – the sound emitted when a rod is struck and the length of that rod – is mirrored in a relation between whole numbers. Uniform rods whose lengths are in simple ratios produce harmonic sounds: 12 to 6, the octave; 12 to 8, the fifth; 8 to 6, the fourth; etc.

The tradition that music possesses powers is very old: Orpheus charmed man and beast when he played his lyre. The Pythagorean school, imbued with mystery, having seen the harmony of music reflected in relations between numbers, was captured. They attempted to order much of what they saw in the world, using the idea of harmonic proportions. Applying this to astronomy, they proposed that the spheres on which planets and stars were fixed rotated so that their speeds of motion were in integer ratios. Thus the planets as they sped through the heavens would emit different notes in harmony. According to Hippolytus,

Pythagoras maintained that the universe sings and is constructed in accordance with harmony, and he was the first to reduce the motion of the seven heavenly bodies to rhythm and song.

We know now, to our sorrow, that the planets sing no song, sacred or profane. They do not move in the harmonies of Pythagoras, on the epicycles of Hipparchus, the circles of Copernicus, or even on the ellipses of Kepler. They move instead on orbits, almost elliptical, but of such complexity that they have no special name and no easy description, according to two rules whose simplicity and elegance are such that Newton could state them in two lines. Because they do the latter rather than the former, we may, if we wish, conclude that the Pythagoreans were wrong; but that they do either is no small wonder. And the Pythagoreans were right in their belief that the entire world could be ordered in some stunningly simple way at its very roots, and that this order could somehow be related to the structure of numbers.

It is a belief that is strong in Plato – of ideal forms behind the flux of phenomena – of rules according to which the planets move, of elementary atomic entities that have the form of solid geometric figures, ultimately triangles, to which they owe their existence. Whether the order is there and we find it (discover the laws of nature) as Plato might have said, or whether the world is so made that we can impose an order on it, is perhaps not relevant. But this belief, daring, naïve, and, considering the almost endless variety of our experience, not at all obvious, that order can be found, has moved scientists from Thales to Kepler to authors in the latest issues of *Physical Review Letters* to create science. For what we call modern science is as much an attitude of mind – a belief in the possibility of a certain kind of understanding – as a set of principles or methods; an attitude of mind such as Albert Einstein expressed after seeing a compass needle at the age of 4 or 5: "Something deeply hidden had to be behind things."

23

The Language of Physics: On the Role of Mathematics in Science

Mathematics is one of the major hurdles that stand between most of us and an understanding of science, especially physics. Those who attempt to write a book on science, accessible to the non-scientist are warned that every equation cuts the, already small, number of potential readers in half. So, can we do without mathematics? Could we express our thoughts in ordinary language? Yes, it might be possible; but it would be extraordinarily cumbersome. Reflect for a moment: music would be possible without musical notation, but imagine giving musicians instructions, for even a simple symphony, using ordinary language without the musician's language: bars, notes etc. – Now all together, flutes, oboes, and strings. This is what physics and much of the rest of science would be without mathematics – possible, but severely crippled.

This essay is based on a chapter originally published in An Introduction to the Structure and Meaning of Physics *in 1968.*

In science, the saying goes, one weighs and measures, one deals in numbers; there is perhaps a suggestion that those elsewhere deal in a currency of less substance. At this point eyes close and minds go to sleep; the decimals unfortunately awaken memories of sultry afternoons in algebra class. It is, of course, true that physics deals with numbers and occasionally with very accurate measurement. One is impressed when an extraordinarily complex calculation of the quantity of magnetism the electron possesses, referred to a certain standard magnet, agrees with experimental observation to ten decimal places. Or when the period of Mercury, calculated according to the laws of Newton, is found to differ from the observed period by ¾ sec out of 8 000 000.[1] The test of a theory, it is thought, is its ability to be in exact numerical agreement with the observations. However, the fact that numbers appear wantonly throughout the body of physics tends to conceal rather than to reveal their significance.

We celebrate Kepler for taking so seriously the discrepancy between one of his proposed orbits and the observations of Tycho Brahe of the planet Mars: a discrepancy of only 8 minutes of arc. This discrepancy finally led him to his famous law: the planets describe elliptical orbits with the sun at one focus of the ellipse. Had Tycho Brahe measured even more accurately, however, it would have been apparent to Kepler that the orbit of Mars was not exactly fitted by an ellipse. Would he then have discarded the elliptical orbit? If he had done so, he would have abandoned one of the most important results in the history of science.

The fact that the planetary orbits are very nearly elliptical is as important a result in this case as if they were exactly elliptical. It is a remarkable regularity; such regularities can be clues to a more precise order or, it is quite possible, could be the only regularities that exist and therefore in some circumstances the limit of our knowledge. We have no a priori guarantee that all the relationships we can discover in the world can be stated with unlimited numerical precision.

Yet, we all know that mathematics is used throughout physics. It is the language of physics. What this means Galileo said very well:

The method which we shall follow in this treatise will be always to make what is said depend on what was said before ... My teachers of mathematics taught me this method.[2]

---

1 The extra 3/4 sec is given by Einstein's theory of gravitation.
2 Galilei, G. (ca. 1590). *On Motion, and On Mechanics: Comprising de Motu.* Madison: University of Wisconsin Press, 1960, p. 50.

"To make what is said depend on what was said before." The result is a structure in which each element has a well-defined meaning and connection to every other element. Pythagoras saw the relation between notes emitted when a metal rod was struck and the length of the rod mirrored in relations between integer numbers. In the years since, physicists have introduced a variety of other mathematical structures in which they have attempted to mirror the world of real objects.

Mathematics, distinct from arithmetic, is a study of the structures that arise out of relations between various well-defined but possibly abstract objects. Much like a game, the rules are defined and all the situations follow as a consequence. In the game of chess, for example, one has a domain of activity – the chessboard; one has the pieces, each of which is allowed certain moves. Every situation that can occur on a chessboard is a consequence. In the game of mathematics, the mathematician begins with certain pieces and a set of rules and then explores the situations or the structure that arises as a consequence of the rules. The rules themselves can be what the mathematician wishes; they need not have correspondence with anything in or out of the world; they need only be consistent. It is the mark of a mathematician, as Bertrand Russell once said, "that he does not know what he is talking about." With inferior rules, however, the structure of a mathematical system becomes trivial, just as when a game is too simple it becomes uninteresting (as in tic-tac-toe, which can always be drawn with the proper moves).

When we think of mathematics, however, we most often think of numbers and perhaps geometry. For it was to the structure of numbers and geometry that mathematicians first turned their attention. The reason is apparent; counting is so primitive that we possibly possessed numbers before we possessed words, or before we could have been identified as human. Animals, more easily taught to count than to understand words, have progressed further than some of our human neighbors. The Watch-andies of Australia count to two, *co-ote-on* (one), *u-tay-re* (two), *booltha* (many), and *bool-tha-bat* (very many). The Guaranis of Brazil adventure further, saying one, two, three, four, innumerable. Was it the ability to count that first separated us from nature? "An honest man," lamented Thoreau, "has hardly need to count more than ten fingers, or, in extreme cases he may add his toes, and lump the rest. I say, let our affairs be as two or three, and not as a hundred or a thousand; instead of a million, count half a dozen, and keep your accounts on your thumbnail."[3]

Possibly one of the earliest motivations for counting was the desire to keep track of a group of animals. At the end of a day of grazing were there as many sheep as there had been at the beginning? A pebble (calculus is Latin for little stone) representing each sheep put in a pile at the beginning of the day could be used in the evening to determine whether the same number of sheep returned. This method, both simple and successful, is even more primitive than counting. It involves only the ability to determine that there are the same number of pebbles as there are sheep, regardless of how many sheep and pebbles are involved. Its success depends also on a fundamental property of the world. Neither pebbles nor sheep disappear into thin air. We know intuitively that the system would not work if we matched the sheep to soap bubbles. Today we might call this the conservation of sheep. With a special provision for antimatter, similar conservation laws exist for nucleons and other particles. If we lived in a world where there were no objects to count, the mathematics of the number system would probably not be among the first to appear; we would not have so primitive and intuitive a knowledge of numbers – and their discovery, rather than being an ordinary jaunt for every child, would require a difficult adventure into a realm of abstract thought.

Geometry arose out of attempts to measure such things as boundaries and areas of land. (The name itself, geometry, means earth measure; much of what we know as geometrical theorems were first observed empirically.) It is easy to see how many

3 Thoreau, H. D. (1854). *Walden.* Boston: Ticknor & Fields.

times a yardstick goes along the side of a field. Of course, the ability to do this presupposes that the world is favorable. The yardstick must not shrink as it is being moved and, for the measurement to make sense, the field itself must maintain its shape to some reasonable degree. Although such suppositions are not always conscious, we should recall that we can measure because we live in a world where we believe we can find yardsticks that do not shrink and fields that do not change their shape. In a liquid world, in which no rigid bodies such as yardsticks exist, it might be much more difficult to measure in a meaningful way.

However, such questions probably did not trouble our ancestors, nor do they usually trouble us. We assume that the lengths and shapes of bodies remain the same when they are moved or turned in space. This seems like an obvious or necessary assumption, but it is not, for we can imagine a world in which such things would not be true. Furthermore (although we do not have to assume this at the moment), we are inclined to believe that the length of the body should remain the same whether it is in motion or not. We shall see later that this assumption will be abandoned.

We thus imagine that we have rods with which we can make triangles, squares, parallel lines, etc.; that we can follow the paths of light rays to produce straight lines of any length; and that we can construct the various geometric figures required. We are willing to assume that rigid bodies exist, that a triangle remains a triangle when we displace it in space, and that when moved it will remain the same triangle. Such properties, when imposed on the world, are not true of necessity but rather something we observe can be true. If we lived in a world without rigid bodies, it would be possible (but not easy) to choose and utilize the same conventions.

Consider the following, possibly outrageous, but perhaps of interest when we consider regions of space as small as those in the interior of a nucleus. Suppose that the space in which we live did not have that quality of rigidity that allows us to mark off distances that remain as marked. Imagine instead that we lived in a world that had some of the qualities of a rubber sheet, continually stretching, twisting, and being otherwise contorted in a completely unpredictable manner. Imagine that living creatures such as ourselves were in this world and that some of them, having nothing better to do, attempted to understand its nature.

Now, clearly it would not be fruitful for them to construct rulers, straight lines, triangles, or to measure distances in their particular world, since the distances between any two points would be continually changing and a measurement at one moment would bear absolutely no relation to a measurement at any other moment. It would be no use to tell a person that it was 2 miles farther from city A to city B than from city B to city C since, having begun the journey, he might find at this particular moment that B was closer to A than to C; it would be better under the circumstances not to mention the matter at all.

In such a world, distance, straight lines, etc., would be meaningless concepts and the mark of a good physicist would be not to introduce them – which does not mean, however, that this world would be without order. Some of the concepts that occur in Euclid's geometry, such as inside and outside, could be defined. Here we see illustrated a dot inside a curve; no matter how the rubber sheet is contorted, the dot will always remain inside the curve.

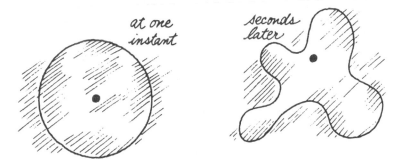

Therefore one could mark off private property or could keep men in prison although one could never say how many acres one owned or whether the prison cell was smaller or larger than the world outside.

From the study of the structures of numbers and geometry, mathematicians have gradually turned their attention to the study of many others. Which structure the mathematician will study is not entirely a different problem from that a painter faces when he asks himself what painting he will paint next. In making such a decision, both the mathematician and the painter have in mind, among other things, the previous history and development of the subject – those areas of investigation that have become overexploited, those that are sterile, and those that seem likely to lead to new and fruitful results.

The underlying program of theoretical physics, both as it is defined by tradition and as it is practiced today, is to find a structure (whose elements are explicitly displayed and logically connected to one another – almost by definition a mathematical structure) some part of which can be put into a correspondence with that domain of the world under analysis. The ultimate aim is to find a single structure that is rich enough so that it can be put into correspondence with all the phenomena of the world. Thus the relations between material things in the real world will be mirrored in the relations between the abstract objects in the mathematical world. Every theory we shall discuss is an attempt of this kind.

Imagine that we are at a baseball game with no knowledge whatsoever of the rules. As we watch, we see many situations, complex or ludicrous. We might begin to see that certain sequences repeat. A batter hits the ball; a fielder catches it, and quite often, but depending on the circumstances, throws the ball toward a position we could call first base. We might after a while have in our minds a certain abstract team in the field consisting of right fielder, center fielder, left fielder, and so on. Any actual team we see will have players with widely varying personalities occupying these positions. In spite of the fact that the size, looks, personalities of the players will vary from team to team, in spite of the fact that the

Giant and Dodger shortstops are different people, as a baseball team these players in their positions will bear certain relations to each other that are independent of their individual characteristics. In that sense, any actual team (one that is in the world) will be a realization of the abstract team we have envisioned in our mind and pictured here.

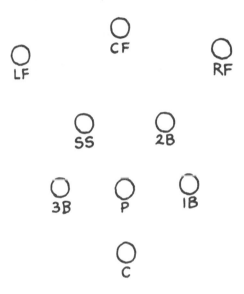

If we watch long enough, we may actually be able to divine the rules according to which the game is being played. But in the game of life there is no set of rules written somewhere against which we can compare our guesses. What systems we invent can be judged by only one criterion – the degree to which they can successfully organize our experience.

At first it would seem that the problem that theoretical physics sets for itself – that of finding a mathematical structure, at least a part of which can be put into a correspondence with the phenomena of the world – would result in nothing more illuminating or less complicated than the world itself, a system governed by rules as numerous as the events observed. If this were so, the enterprise would not be particularly rewarding. However, it has turned out that we can find mathematical structures that can be put into a correspondence with the events of the world, such that the rules underlying the working of the structures are simple and few in number. This could hardly have been anticipated and is perhaps the most remarkable fact about our world.

And if so inclined, one can identify these few rules according to which everything might be thought to move as being similar in spirit to those ubiquitous ideas of Plato, those ideas that to Plato meant what numbers meant to Pythagoras: those essences that lay behind the working of reality. A remarkable example is the structure assembled by Euclid to organize our experience of space.

"There seems to be a pattern of some kind, but what I can't figure out is *why*"
(Drawing by O'Brian, ©1967 *The New Yorker Magazine*, Inc.)

24

The Structure of Space

Euclid's Elements *provides one of the earliest and best examples of a system in which complex relationships follow from a small set of very simple assumptions. But is space Euclidian? Is it curved? Does light travel in straight lines? What do such questions mean?*

This essay is based on a chapter originally published in An Introduction to the Structure and Meaning of Physics *in 1968.*

Euclid's Elements

Geometry, like Latin, with generations of repetition has become synonymous with the trials of adolescence and evidence of the inhumanity of adults to their young. Much has passed since Plato had inscribed over the doors of his Academy, "Let no man ignorant of geometry enter here," or since Edna St. Vincent Millay wrote, "Euclid alone has looked on beauty bare."[1] What there is in the *Elements* of Euclid that made them the model for the science of Galileo and Newton, and for the philosophy of Descartes. Why they provide a gem-like example of a mathematical system and of a physical theory remains a mystery to numberless students for whom Euclid evokes only a memory of pain.

In a world beset by uncertainty, the demonstrations of geometry at one time seemed a model for what a sure argument should be. A

---

1 St. Millay, E. (1922). *The Ballad of the Harp Weaver*, New York: printed for Frank Shay (Flying Cloud Press).

dispute in the marketplace begins obscurely, and ends in turmoil. In political arguments opinion sways first to one side and then to the other, fluttering like a butterfly, finding no place secure enough to rest. But in geometry, once one has granted the postulates, the entire structure follows with an inevitability that cries to be emulated. In the words of a typical geometry text, "Every proof consists of a number of statements, each of which is supported by a definite reason." With such a method it would appear one could achieve certainty.

Of course, this kind of certainty had been available elsewhere, for example, in the syllogism of Aristotle. We would grant as certain that if all men are mortal and if Socrates is a man, then Socrates is mortal. But the syllogism, while certain, contains no surprise. Having granted the first two statements, the third comes as no revelation. But having asserted Euclid's five postulates beginning "Let it be granted that it is possible to (1) draw a straight line from any point to any point," and concluding with the famous postulate concerning parallel lines, there follows the not-obvious consequence that the sum of the angles of a triangle is 180 degrees, or that for a right triangle the sum of the squares of the sides is equal to the square of the hypotenuse. It is this element of surprise in Euclid that is perhaps his most captivating achievement. For in Euclid's system it appears that we can achieve certainty without triviality.

In considering what it was that Euclid did, we should recall that most of the relations that occur in his geometry were not derived first by him but had been observed or discovered previously, possibly in the process of land measurement. Rather than the beginning, Euclid's *Elements* are the climax of possibly a thousand years of research in geometry. Others before him had established propositions and chains of propositions. It was known long before Euclid's time that the sum of the angles of a triangle was 180 degrees. And, of course, it was known by Pythagoras at least 300 years before Euclid that the sum of the squares of the sides of a right triangle is equal to the square of the hypotenuse.

The accomplishment of the geometer, as epitomized by Euclid, was to show that all these diverse relations followed from certain very simple assumptions. Upon reflection we realize that the assumptions then contain the entire structure of the geometry. But what Euclid revealed was just this structure, just the interrelations between one proposition and another and between the postulates and all the propositions.

He begins with twenty-three definitions in which he attempts to describe what the objects are that he is talking about. The effort is not completely successful. He says, for example (definition 1), "A point is that which has no part." Or (definition 2), "A line is breadthless length." And in his fourth definition, whose meaning has eluded geometers ever since, he says, "A straight line is a line which lies evenly with the points on itself." (This might mean, "A straight line is one that is not curved.")

And so on, to definition 23, "Parallel straight lines are straight lines which, being in the same plane, and being produced indefinitely in both directions, do not meet one another in either direction." If all these are not entirely illuminating, it is not the fault of the reader, for it has taken mathematicians two millennia to understand what they mean. Euclid then follows by asking us to agree to the five postulates, saying: [Let the following be granted:] (1) "that it is possible to draw a straight line from any point to any other point;" etc. Then the five axioms: (1) things which are equal to the same thing are also equal to one another; (2) if equals be added to equals, the wholes are equal; and so on to (5) the whole is greater than the part. These axioms, or common notions, are distinguished from the postulates in that they represent an agreement as to what the language (such things as "equals," or "added," or "subtracted") is to mean. They are common presumably to all systems, in contrast to the postulates, which are specific to his geometry (a distinction introduced by Aristotle). Since he asks us to grant the postulates, presumably it would be possible to deny them and to grant something else.

From these rules and definitions the structure is built, as the geometry text says, "every statement being justified by an axiom or postulate, or a previously proved theorem." The sum of the angles of a triangle is 180 degrees, the sum of the squares of the sides of a right triangle is equal to the square of the hypotenuse; these and all the other theorems of geometry follow with a certainty that is unchallengeable. It is this certainty, attainable in geometry, that has raised the hopes of philosophers and others that similar certainty could be achieved elsewhere. Descartes, for example:

The long chains of simple and easy reasonings by means of which geometers are accustomed to reach the conclusions of their most difficult demonstrations, had led me to imagine that all things, to the knowledge of which man is competent, are mutually connected in the same way...?[2]

But though the structure of geometry seems clear, there have been many opinions about the meaning of the definitions and the postulates. Are they what Descartes would call "something we know clearly and distinctly, and thus cannot be doubted?"[3] Are they what Aristotle would describe as "something which is intelligible, or intrinsically knowable?" Or are they, as Immanuel Kant stated, "propositions which are thoroughly recognized as absolutely certain ... and yet as independent of experience?"[4] If not, in what sense can we believe them? Or in what sense can they be denied?

Is space Euclidean?

There are few questions that seem as utterly incomprehensible as: "Is space curved?", "Do parallel lines meet at infinity?" and so on. And when questions are so baffling, the suspicion, often true, is

2 Descartes, R. (1637). *Discourse on the Method of Rightly Conducting the Reason and Seeking Truth in the Sciences*, transl. John Veitch, La Salle, IL: Open Court, 1949, p. 19.
3 Ibid.
4 Kant, I. (1783). *Prolegomena to Any Future Metaphysics*.

that they are so formulated as to be contradictory. One can be baffled endlessly considering the problem of an immovable wall and an irresistible force, because clearly (or possibly not clearly) an immovable wall and an irresistible force are self-contradictory in the same universe. The source of much of the difficulty in questions about space lies in confusion about the meaning of the definitions and the postulates of geometry. And since geometry may be considered either a model of a mathematical or a physical system, some thought about the meaning and interpretation of these definitions and postulates can be very useful.

Let us consider such a simple thing as a straight line. What is it? We listen to Euclid (definition 4): "A straight line is a line which lies evenly with the points on itself." Further, the first postulate: [Let it be granted that it is possible] "to draw a straight line from any point to any point," which we take to mean, "Let us assume that one, and only one, straight line can be drawn from any point to any other point." This would seem to preclude the possibility shown in Figure 24.1. In the words of Euclid or his immediate successors, "Two straight lines cannot enclose a space." Any opinion poll not restricted to logicians would reveal general agreement with this, along with the observation that the lines in Figure 24.1 are obviously curved.

Yet suppose what we thought to be straight lines actually were curves drawn, for example, on the surface of a sphere, as shown. Such "straight lines" would have no difficulty in enclosing a space. But what does it mean to say that what we "think" are straight lines are "actually" curves? We might first ask what in the physical world do we think is a straight line? Or, put in another way, "How shall we actually construct a straight line?"

Several methods suggest themselves. We might, for example, stretch a string as tightly as possible and say, "That is a straight line." Or perhaps we might project a ray of light and say, "That is a straight line." But as soon as we do this, we are making, without analysis, the critical assumption that those physical objects, rays

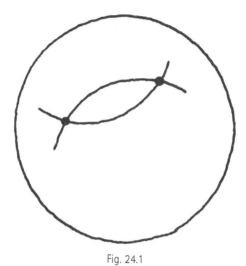

Fig. 24.1

of light or stretched strings, which are in the real world, have the properties of the Euclidian straight line.

Yet we might imagine that we lived in a world in which light rays or stretched strings followed curved paths. To say that a straight line as defined by a light ray or a stretched string has Euclidian properties is an assertion about the world in which we live; an assertion that may or may not be true; an assertion whose truth can only be verified by actual observation, because we can just as easily imagine a world in which light rays travel in "curved" as in "straight" lines.

Fig. 24.2

The identification of light rays with straight lines lies so deep in our mentality that we are sometimes led to incorrect conclusions because of it. When we place an oar in water and look at it, it appears bent (see Figure 24.2). The eye and the mind prefer to see the oar bent when it is not, rather than to admit that light in this case does not travel in a straight line.

The difficulty goes all the way back to Euclid. Consider again his attempt to define a straight line. He says (definition 4): "A straight line is a line which lies evenly with the points on itself." But what does that mean? (The meaning of this definition has been a subject of conjecture by geometers ever since.) Thomas Heath, in his translation of Euclid's *Elements*, suggests that Euclid was aware of Plato's definition of a straight line, which was: "Straight is whatever has its middle in front of both its ends" – which is to say that to the eye, placed properly, a straight line appears to be a point. But of course this will be so only if light travels in straight lines. Euclid revealed his genius when he realized the difficulty in Plato's definition and tried to avoid it by attempting to divorce the notion of a straight line from any physical phenomenon. Having written a treatise on optics, he knew that light did not always travel in straight lines, so that the path a ray of light followed in fact did not always satisfy his postulates.

What then can we make of this? Imagine that we lived in a world in which light did not travel in straight lines. For example, suppose we lived on the surface of a sphere (which we do), but suppose we were constrained so that nothing could ever get off, all the balls we threw, strings we stretched, rays of light, etc. always moved along the surface. In such a situation we could never construct an object that had the properties of Euclid's straight line. Would we then say that parallel lines meet at infinity, that our space was curved and non-Euclidian? Although it possibly would be more fruitful to treat it that way, we would not have to. But we shall return to this later.

Euclidean geometry as a mathematical system

Geometry was formulated as a completely mathematical or logical system by David Hilbert at the end of the nineteenth century. Many of the Euclidian presuppositions, such as those involving congruence, etc., do in fact involve the physical properties of space. The difficulty of disentangling geometry as mathematics from geometry as a statement about the physical world cannot be made more clear than by the fact that it took so long to achieve the separation.

We consider geometry as a mathematical system by focusing our attention on the definitions of the primitive objects. On the whole they are confusing. "A point is that which has no part." Does that help us understand what a point is? "A line is breadthless length." Is that illuminating? From the point of view of a mathematical system such definitions are not only not illuminating, but not necessary. For what a point or a line is, is supremely irrelevant. As a mathematical system, geometry concerns itself with the primitive and undefined entities called points, lines, etc. that do not have to be defined at all. (Currently, they would be called "undefined terms.") What matters is that certain relations should exist between them,

Fig. 24.3

such as "Through any two points one may draw one, and only one, straight line." Given the relations between the undefined terms, further relations can be proved; these are the theorems. The structure of such a system is independent of what the points or lines actually are.

This is not very difficult to understand. Consider the game of chess. There are certain given pieces – castles, bishops, knights, queen, king, and pawns. The pieces obey rules; each of them is allowed to move in a certain way; the game begins when each is placed in a specified position on the board. Now, although we all know roughly what the knight looks like (it's something like a horse), or what the queen looks like, or the king – and it is useful to know this, otherwise we might forget as we looked at them on the board – it is perfectly clear that the game of chess is the same, no matter how the knight appears – traditional and ornate, modern and abstract, or utilitarian and inexpensive (see Figure 24.3). If, to teach children chess, we begin by saying that a knight is a rider on a horse, the king wears a crown, the castle has a turret, etc., it is clear that we do this for purposes of pedagogy, identification, or familiarity. Because what is relevant for the game of chess is only

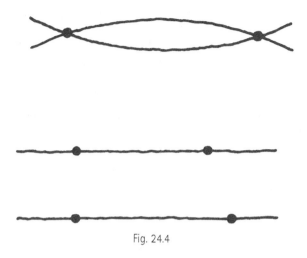

Fig. 24.4

the fact that pieces stand in a certain relation to each other – that each piece is allowed certain moves. All the situations that follow are consequences of these rules. They depend not at all on whether the set is made of ivory, steel, or of sticks put into corks.

Viewed as a mathematical system, or as a model of all mathematical systems, the situation in Euclidian geometry is precisely the same. Whether a line is breadthless length, whether it is the idealization of a stick made infinitely thin, whether it is to be thought of as a stretched string, or as a light ray, is a matter of complete indifference. Such realizations may aid us in our visualization or in our intuitive feeling for the property of these objects, but they also confuse us. For what is important to the geometry is that whatever it is that we have called a line, and whatever it is that we have called a point, satisfy the postulates of geometry. If they look like the first figure, then clearly they will not satisfy the postulates; if they look like the one below it, then possibly they will (see Figure 24.4).

Thus, as a mathematical system, geometry has its pieces: lines, points, etc. It has its rules: the postulates. And using the rules to manipulate the pieces or to construct other pieces (for example, triangles), all the theorems follow, all tied one to the other in a marvelous and subtle structure, every one of whose elements is connected in a clear and definitive way to all the others. Our concern with this particular system (the reason we are interested in straight lines, points, triangles, and so on) of course originates in the possibility of associating it with certain real objects of the world. And surely, the first notions of lines and points arose as abstractions from objects almost with breadthless length, and objects almost with no parts. But as a mathematical system, geometry stands independent of these associations. What counts are the relationships between those finally undefined objects; what is of interest is the structure itself.

Euclidean geometry as a physical system

Treated as above, Euclidian geometry says absolutely nothing about the world. Such a structure of theorems might be self-consistent and true, or contradictory and false, no matter what the nature of the world is; in just the same way as it is not possible to checkmate the black king with a white knight and king, independent of whether the knight and king are made of ivory, in a traditional style, or of steel in a modern style.

Yet, that the statements of geometry do have something to do with the world could not be clearer. Consider the triangle shown. If we cut out the angles A, B, and C, and place their points together, we find that they make a straight angle, or a straight line.

We find this invariably, over and over again, for any triangle we cut out carefully from a piece of cardboard. Now, it is clearly a statement about the physical world to say that if we draw a triangle on a piece of cardboard, cut out the angles, place them together, they add up to a straight line, a straight line being something that lies parallel to the edge of a ruler. There must be a sense, then, in which the theorem "The sum of the angles of a triangle equals 180 degrees" is a statement about the world (see Figure 24.5).

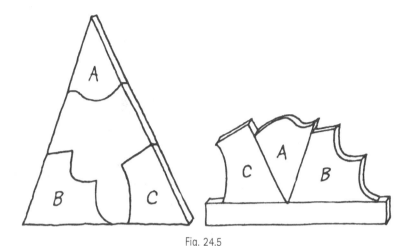

Fig. 24.5

We can make geometry a statement about the world by indicating how in the world we are to construct those undefined quantities of the theory, those quantities that Euclid tried bravely, but not successfully, to define. We must state what it is we are to mean by a point, what it is we are to mean by a line. This, as Newton much later was to shrewdly realize, was not a problem of geometry itself. He says:

Geometry does not teach us to draw these lines, but requires them to be drawn, for it requires that the learner should first be taught to describe these accurately before he enters upon geometry, then it shows how by these operations problems may be solved. To describe right lines [straight lines] and circles are problems, but not geometrical problems.[5]

The construction of a point or a straight line is not a geometric problem. Geometry presupposes that one is given objects – points and lines – that have the properties required by the

5 Newton, Sir Isaac (1687). *Mathematical Principles of Natural Philosophy and His System of the World*, vol. 1, transl. Andrew Motte, Berkeley: University of California Press, 1962, p. xvii.

Fig. 24.6

postulates. If, in the real world, we can exhibit such objects, then, since they satisfy the rules, they will also exhibit the properties of the theorems.

Now, when one speaks of any drawn line as an approximation to a straight line, what one means is that any drawn line, to the extent that it has small irregularities in it, does not exactly satisfy the postulates, etc. But there is a sense, which we understand among ourselves, in which a drawn line satisfies the postulates as well as we wish to make it satisfy them.

So when we interpret geometry as a physical system, what we mean is that we can construct objects such as the top of Figure 24.6, which we call points, and other objects such as the one at the bottom of Figure 24.6 that we call lines, and that the objects thus constructed satisfy the postulates. For example, only one line can be drawn between two points. Having done this, our geometry has now taken on the aspect of a physical theory. For among the objects thus drawn will exist the relations that occur in the geometry – the sum of the angles of a triangle will be 180 degrees, and the sum of the squares of the sides of a right triangle will be equal to the square of the hypotenuse.

The most delicate problem is the interpretation, the association of the objects in the physical world with the undefined objects in the mathematical structure – the assertion that the object we display has the properties of a straight line or that the object we construct has the properties of a triangle. In the case of geometry, it is fairly straightforward. But in physical theories of greater complexity it is the continual association of the abstract objects of the mathematical world with the corresponding objects of the real world that present the most difficult conceptual problem.

Euclidean geometry as a convention

In what sense, then, can space be non-Euclidian? If Euclid had done nothing else, he would be remembered for his introduction of the fifth postulate (later called the "parallel axiom"). It reads: "…if a straight line falling on two straight lines makes the interior angles on the same side less than two right angles, the two straight lines, if produced indefinitely, meet on that side on which are the angles less than the two right angles." Euclid felt that this postulate was necessary, for example, to prove the theorem that the sum of the angles of a triangle is equal to 180 degrees. Whether or not it is became a subject of contention for centuries afterward. Geometers from Ptolemy and Proclus through mathematicians of the nineteenth century attempted to show that this fifth postulate itself was a consequence of the other four.

The problem and the difficulty is that our minds tend to be riveted to a space that is itself Euclidian – to a space made up of physical rods and physical points which themselves have the properties of the parallel axiom. It was not until the nineteenth century, when

Lobachevski and Bolyai showed that one could construct a consistent geometry in which more than one line could be drawn parallel to another through a point, that the independence of the fifth postulate became clear.

By the end of the nineteenth century, two alternatives to Euclidian geometry had been developed. The first, Riemannian geometry, was essentially the geometry of a surface of a sphere. In this case, no lines can be drawn parallel to a given line through a point. The curves defined as straight lines meet at poles. The second was the geometry of Lobachevski and Bolyai, in which many lines could be drawn parallel to a given line.

Now, how is one to decide whether the space in which we live is Euclidian or not? Perhaps the easiest way is to take an immediate consequence of the parallel axiom. For Riemannian geometry – the geometry of the surface of a sphere – the sum of the angles of a triangle is larger than 180 degrees and grows larger and more different from 180 degrees as the triangle grows larger. For Euclidian geometry the sum is just 180 degrees, and for the geometry of Lobachevski it is less. All one has to do, then, is to take what are called straight lines in the real world and to measure (presumably taking a large triangle) whether the sum of the angles of this triangle is 180 degrees or more or less. Such an experiment was suggested by Schweikart. And Gauss made an attempt to measure directly with surveying equipment by ordinary triangulation whether or not the sum of the angles of a triangle was 180 degrees. A triangle formed by three mountains of the order of 100 km apart did not enable him to detect any deviation from 180 degrees.

Measurements to determine whether or not light rays define Euclidian straight lines have been made using fixed stars as vertices of triangles. Lobachevski, for example, used a triangle whose base was the diameter of the Earth's orbit and whose apex was the star Sirius. Again, no deviations from 180 degrees were found. He thus wrote:

However it be, the new geometry whose foundations are laid in this work, though without application to nature, can nevertheless be the object of our imagination; though not used in real measurements it opens a new field for the application of geometry, to analysis and vice versa.[6]

Even measurements of the triangle bounded by the diameter of the Earth's orbit and the star Sirius are relatively small compared to triangles that span the entire known universe. And it is possible that the measurement of such a very large triangle, whose sides were determined by light rays, would give us angles whose sum was larger or smaller than 180 degrees. Suppose we triangulate light paths going from one end of the universe to the other and find that the sum of the angles of a triangle is different from 180 degrees. Would we then conclude that space is non-Euclidian? Not necessarily. It now would depend entirely on the point of view we wished to take.

Again, imagine that we were constrained to the two-dimensional surface of a sphere and could not leave it. And imagine further that such things as light rays or stretched strings followed the surface of this sphere to form great circles. If we triangulated on the surface of this sphere using light rays, for example, we would find that the sum of the angles of a triangle was larger than 180 degrees. Now, as pointed out by Poincaré, we could take two points of view. We could say that we lived in a space that was non-Euclidian – that is to say, in a space in which one could draw no line parallel to a straight line. But we could equally well say that we had not chosen our straight lines properly – that is to say, that what we thought were straight lines, light rays, stretched strings, and so on, were not in fact straight lines, because they did not have the Euclidian properties.

6 Lobachevski, N. (1840). *Neue Anfangsgrunde der Geometrie*, p. 24 and quoted from Jammer, M. (1954). *Concepts of Space*, Cambridge: Harvard University Press, p. 24.

Looking at the surface of the sphere from the outside, which is essentially embedding it in a three-dimensional space that is Euclidian, we see immediately that we are dealing with curves on the surface of the sphere, and not "real" straight lines. And so we could take the point of view that space, "real" space, is Euclidian, but it is our fortune, or misfortune, to be living on the surface of a sphere, a surface that prevents us from easily drawing those objects (straight lines) that have the Euclidian properties and satisfy the Euclidian postulates.

Thus it becomes a matter of convention whether or not we consider space Euclidian. If we decide, for example, that light rays should travel in straight lines and then in a measurement we find that triangles formed with light rays do not have angles that add up to 180 degrees, we can reject the light rays as straight lines and look for something else. Clearly, we can always do this; and it is this that is the source of the major part of the confusion about whether space is Euclidian or not.

When a new physical theory is interpreted as saying that space is not Euclidian, what could be meant is that in the space in which we live such things as light rays travel in paths that do not have the properties of the Euclidian straight line. What baffles the imagination is (if it is true that light rays do not travel in straight lines, and one might be willing to admit that) the attempt to embed whatever it is the light ray is traveling in into a space in which one can draw some kind of a straight line. Thus if the light ray is traveling on the surface of a sphere, we imagine the sphere embedded in a three-dimensional Euclidian space. Thus "real" space is Euclidian. It just happens that we live on the surface of a sphere.

As long as the argument is about convention, neither side can win. But the physical theory that states that space is non-Euclidian is stating something more than convention. It is stating that the convention of looking at space as Euclidian is not a fruitful one. For example, if light rays travel as though they were on the surface

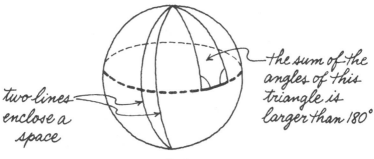

two lines enclose a space

the sum of the angles of this triangle is larger than 180°

Fig. 24.7

of a sphere, if stretched strings bend as though they were on the surface of a sphere, if a particle released in uniform motion travels as though it was on the surface of a sphere, if in fact we are incapable of constructing in any way those objects that have the properties of Euclidian straight lines, of what use is it to us to assert that our space is Euclidian? We can do this if we wish. But it is a Euclidian space in which we never can construct the objects that have the necessary Euclidian properties. If we insist, we can call the "real" space Euclidian. However, it becomes an encumbrance increasingly less fruitful. In the end, if we are constrained to the surface of a sphere, it is easier (but not necessary) to speak of our space as a non-Euclidian space, one that has the properties of the surface of a sphere, rather than speaking of our space as Euclidian, although a Euclidian space in which we are prevented ever from constructing a straight line.

If we were constrained to the surface of a sphere, if light rays, etc. moved along paths as shown, then two "lines" could enclose a space, the sum of the angles of a "triangle" would be larger than 180 degrees, etc. (see Figure 24.7). We could take the point of view that space was "really" Euclidian, but unfortunately light rays did not travel exactly in straight lines, deviating from "exact straightness" by an amount depending upon the length of the path. The

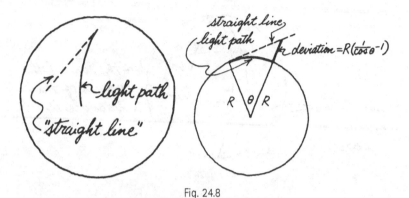

Fig. 24.8

amount of deviation could be expressed in a formula:

$$\text{deviation} = R \left(\frac{1}{\cos \theta} - 1 \right). \tag{24.1}$$

There certainly would arise an occasion when we could remind a colleague: "It's well known that light rays don't travel in straight lines (nothing does). To obtain a straight line we follow a light path and then correct" (see Figure 24.8).

The major difference in point of view between Newton's and Einstein's theory of gravitation (general theory of relativity) lies in their convention concerning the geometry of space and time. In the Newtonian theory space is Euclidian, and particles that move on curved paths do so because of forces. In general relativity, space-time is non-Euclidian, and particles always move in such a way that they traverse the shortest distance between any two points, given the constraints of the space. The two points of view, although there are important differences, lead in most cases to almost identical results – an illustration of the extent to which the choice is a matter of convention. Which convention we choose is determined essentially and ultimately by its fruitfulness. The convention is created by the mind, and the statement about the world is that a certain convention enables us to organize it successfully.

Poincaré believed that no convention would be more convenient than that of Euclid; only fifteen years later Einstein proposed the general theory of relativity, in which he substituted a non-Euclidian geometry. But perhaps one should not say "more convenient." General relativity, although startling in its beauty and elegance, has never been a "convenient" theory. And most attempts to use its non-Euclidian geometry have essentially treated it as a deviation from the Euclidian space because, in comparison with any other, the Euclidian space is easier to use.[7]

7 Recent measurements suggest that space is Euclidian, that is, light travels in straight lines and the sum of the angles of a triangle is 180 degrees.

25

Superconductivity and Other Insoluble Problems

Before they are solved, scientific problems often seem very difficult, perhaps impossible. But once they have been solved, the solutions sometimes seem obvious, even inevitable. But are there some scientific problems that are genuinely insoluble? If so, what are they? And how would we know?

On this fiftieth anniversary of BCS (Superconductivity, not the Bowl Championship Series) it's hard to recapture the very great difficulty (even conjectured insolubility) this problem appeared to present just half a century ago. Many of the greatest names in physics had tried or were trying their hand – Einstein, Bohr, Heisenberg, and Feynman, among many others. The contrast between then (impossible) and now (safely ensconced in textbooks) leads me to reflect a bit about problems, past and present, thought to be insoluble.

Problems are thought to be insoluble for different reasons: it may be too soon, as Newton's attempts to understand the properties of matter, or Einstein's 1922 attempt to construct a theory of superconductivity before a quantum theory of metals was in place. It may be that existing science does not contain the solution – a new postulate or law of nature is required, as Max Planck's 1901 quantum hypothesis. Or might it sometimes be that no solution is possible? Put another way: Are there limits to what we can understand? Are there limits to science as we do it? Are some questions inaccessible to human intellect?

The long and very imposing list of physicists who had tried their hand at superconductivity should have been enough to give me pause. But, fortunately, I was relatively unaware of their many unsuccessful attempts when, in the spring of 1955, John Bardeen invited me to join him to work on this problem. As is now very well known, our collaboration resulted in what is universally referred to as the BCS theory, which, after six months of non-stop calculation, we submitted for publication in July 1957. We knew that we had solved the problem, but could not, at that time, appreciate all of the ramifications. It was as though we set out to build a car and along the way invented the wheel.

Remarkably, with a few exceptions, acceptance of our theory was almost immediate. There were some complaints – lack of gauge invariance was one. We had discussed and understood this but, wisely, had not wasted too much time on it. There were also some regrets: one rather well-known low-temperature physicist (possibly hoping for a new law of nature) expressed his disappointment that "such a striking phenomenon as superconductivity [was]...nothing more exciting than a footling small interaction between atoms and lattice vibrations." On a personal level, what was most astonishing to me was how quickly we had gone from impossibly difficult to obvious.

After a seminar I gave in the early 1960s trying to demonstrate how the pairing state comes from the full many-body problem, a young man asked me "Why are you working on this problem? Everyone knows that there is a pair condensation."

It is difficult to imagine today that superconductivity was once regarded by some as an insoluble problem. Felix Bloch said, probably as somewhat of a joke, "Every theory of superconductivity can be disproved." But there is, and has been, no shortage of, supposedly, insoluble scientific problems.

In the sixteenth century, Gilbert wrote:

...Alexander Aphrodiseus incorrectly declares the question of amber to be unsolvable, because that amber does attract chaff, yet not the leaves of basil.[1]

But there are current problems that are occasionally regarded as impossibly difficult, even insoluble. One is at the forefront of physics: the problem of the values of the parameters that are required for the very successful standard model. There are currently twenty-six of these. Is there a theory that will give us the values of these standard model parameters? Among the discouraging facts often cited are the huge differences between the masses of the "elementary" objects: top quark mass of 1.74×10^{11} electron volts, electron neutrino mass equal to or smaller than 2.2 electron volts, three force coupling constants, the fine structure constant etc. I will not list them all.[2]

One remarkable fact about these parameters is what is called their fine-tuning – that is, small differences of the standard model parameters would result in a universe dramatically different from the one we live in. If, for example, the down quark were lighter than the up quark the proton would decay, hydrogen would not be stable so that, presumably, neither would we – a universe in which life as we know it would not be possible.

Now life as we know it is not possible on newly demoted Pluto, the moons of Jupiter, the interior of the sun or, perhaps, for that matter in some parts of Rhode Island. We have, in fact, not yet discovered any other habitat where life exists. Another universe might have more such habitats, or none at all.

Whether this "fine-tuning" is someone's design in order that we might come about, a statistical accident, as is suggested in what is called the "Anthropic Principle" or the consequence of an, as yet unknown, deep underlying theory is something we do not know.

1 Gilbert, W. (1600). *De magnete*, transl. R. Fleury Mottelay as *On the Lodestone and Magnetic Bodies*, New York: John Wiley, 1893.
2 Recent calculations yield good values of proton and quark masses.

But let us project ourselves backward to a problem facing physicists at the beginning of the twentieth century: the radiation emitted by a single atom: hydrogen. Among the hundreds of spectral lines emitted by hydrogen only a minute few are visible to the human eye. Who would have believed that we would ever have a theory that could predict every one? Yet we can now calculate these wavelengths, that range from X-rays, wavelengths smaller than 10^{-8} cm (if one tosses in hydrogen-like muon–nucleus systems) to microwaves, of wavelength 28.4 cm (the Lamb shift) – a minute splitting of two of the hydrogen energy levels due to quantum electrodynamic interactions, a span of at least ten orders of magnitude. Who would have been bold enough to predict that within fifty years we would have a theory that yielded all of these with incredible precision and, in the case of one fundamental quantity (the magnetic moment of the electron), the very latest comparison of theory with experiment is in agreement to at least ten decimal places.

Or, projecting ourselves backward some 10×50 years, consider the state of pre-Copernican astronomy (see Figure 25.1). By the time of Copernicus, astronomers were using up to eighty epicycles to fit the planetary data. Could the medieval astronomer have imagined that all of the complexities of the planetary motions would follow as consequences of two postulates?

$$F = ma \qquad (25.1)$$

and

$$F = G \frac{M_1 M_2}{R^2}. \qquad (25.2)$$

A current, seemingly, insoluble problem might be: What happened before the Big Bang?

However, consider a physical theory that resulted in the following series:

$$1 + x + x^2 + x^3 + \cdots. \qquad (25.3)$$

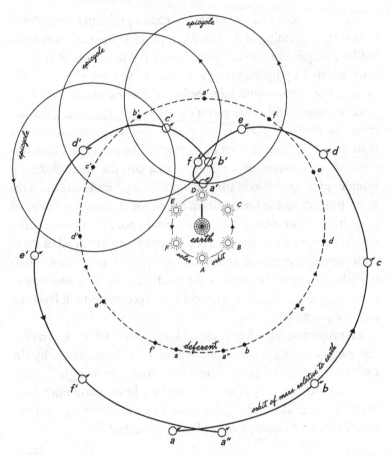

Fig. 25.1 In the Ptolemaic system, epicycles were postulated in order to explain the observed retrograde motion of planets, while also preserving the geocentric model of the universe. Epicycles were likewise used in the Copernican system in order to fit the observed data and maintain circular orbits.

The radius of convergence of this series is 1. We might, therefore say that theory could never be extended to values larger than $x = 1$. But, of course, for $x < 1$ this series is the expansion of the function

$$\frac{1}{1-x} = 1 + x + x^2 + x^3 + \cdots. \qquad (25.4)$$

Thus we can analytically continue beyond the singularity. Might such a method be used to tell us what happened before the Big Bang?

Of course, all of this in no way assures us that a theoretical structure that gives us the parameters of the standard model can be invented. We cannot rule out the possibility that our universe with its own particular life friendly parameters is one of 10^{500} existing and possibly inaccessible universes. Or, that the view that sentient creatures, such as ourselves, probe this universe because it allows our existence is wrong (or not even wrong as Pauli might have said). But, as we have seen for superconductivity, what seems to be impossible one day becomes ordinary the next.

I am reminded of conversations that took place among young members of the Princeton Institute for Advanced Study almost sixty years ago. We guys (all guys at that time) would sit around bemoaning the fact that Einstein, Heisenberg, Schrödinger, and Feynman had solved all of the easy problems, leaving the hard ones for us. In the time that has passed most of those then so-called hard problems have been solved – some of them by us. Very likely today's young members bemoan the fact that we solved the easy problems leaving the hard ones for them.

But it is legitimate to ask simply: in spite of our remarkable, even astonishing, success, are there limits? Are some things intrinsically unknowable – perhaps, the state of the universe before the Big Bang? We don't know the answer to this question. But, before we leap to conclusions, let us recall Auguste Compte's example of what is intrinsically unknowable: the chemical composition of the stars.

Or, perhaps as some believe, we have been put here to discover all there is. As in the following conversation that, we are told, took place between Abraham and God.

God says to Abraham:

Without Me you would not exist.

Abraham replies:

> Without me, no one would know that You exist.

We can be confident that physics and science 2×50 (not to speak of 10×50) years from now will be as remarkably different from what we have at present as what we have now is from what we had 2×50 years ago. We can also be reasonably confident that much of what is currently thought to be impossibly difficult or even insoluble will be enshrined in textbooks as totally ordinary. ("Everyone knows that there is a pair condensation.") But not to fear. Other insoluble questions will have taken their place.

26

From Gravity and Light to Consciousness: Does Science Have Limits?

In the last nine hundred years of scholarship and research a remarkable body of work has been created. But can this continue indefinitely? In spite of our great progress, we may ask if science has limits? And if science has limits, what are they?

The essay is based on a lecture given at a symposium celebrating the 900th anniversary of Bologna University in 1988.

It is with pride and pleasure that I speak to you on this nine-hundred-year celebration, this anniversary of the signing of your Magna Carta: nine hundred years of intellectual activity, nine hundred years of continuous university existence. Of course, during these nine hundred years we expect a certain sufficiency of bureaucracy, excess of tenured faculty, occasional bad teaching, and some sleepy students. But these nine hundred years are also (I'll try not to be too effusive) a triumph of our all too human intellect – possibly alone in the universe – our struggle to understand, our struggle against brute nature and dark superstition. Is it out of place to paraphrase that cigar-smoking, whiskey-drinking, charmer? "Never have so many (all human beings) owed so much (of what we enjoy) to so few (ladies and gentlemen, that is us)."

You are probably thinking, "Can he be serious?" Yes, this is a grand occasion but doesn't he read the papers? Radiation, plutonium, ozone, and new viruses. Hasn't he heard of the dangers as well as the limits of science? Yes, I do read the papers. In recent months, I have read at least one article too many about the evils and the limits of science. I cannot avoid the latest intellectual fashions: catastrophe theory, chaos, complexity, computability. As in antiquity, when the gods are jealous, they take revenge; for the unsurpassed gifts we have offered to humanity: twentieth-century fire – electricity, antibiotics, polio vaccine to mention a few – they try to shackle us to be savaged by eagles of ideology, deconstruction, and revision. We await Heracles.

Do I dare say that much of this talk comes from certain computer scientists and philosophers who have never done any science nor seem to have any concept of what it means to do it? They are not the only ones. Sociologists and political reformers reward us for our very hard work and occasional success by accusing us of irrelevance or elitism – criticisms I have never heard addressed to sports such as football or downhill skiing. There the unquestioned, socially useful and politically correct, criteria are abilities to kick more goals or to make descents measured in parts of seconds.

So, true to my reputation, somewhat to the right of Nero, if not Attila, let me attempt in my own way to celebrate this occasion. To discuss possible limits of our human intellect, I would like to illustrate its actual workings, taking examples from the past as well as those for the future. Since I have spent my professional life in the sciences, you will forgive me if I take illustrations from the fields I know. I might attempt to offer a view of Dante's *Divine Comedy* (for we would rather talk about what we would like to know than what we are supposed to know). But I will be prudent and forbear. I will conjure up three ghosts (my respects to Dickens); the ghosts of Science Past, Science Present, and Science Future.

The first may not be familiar as an example of science but you will recognize the opening lines.

Nel mezzo del cammin di nostra vita
mi retrovai per una selva oscura
che la diritta via era smarrita[1]

Ah! Dante you say? Why not? He did visit Bologna. The great "Italian" poet, moralist political commentator, but scientist? In this intimate gathering – and just between us – I do admit that Dante's particular interest was the appropriate eternal reward for his various colleagues, friends, acquaintances as well as politicians and popes. However, he was careful to place them in a meticulously designed universe totally consistent with the Aristotelian conception. And since I have invoked his name, I can't resist showing you one of Gustave Doré's illustrations of his remarkable journey, the ninth circle of Hell: three-faced Satan, frozen hip deep in a lake of ice, where, embedded for eternity to a depth appropriate to their crimes, are some of Earth's greatest sinners. On the right, if you look closely, you will see two figures: Dante and his personal guide, the Latin poet Virgil (see Figure 26.1).

Dante, as we have just mentioned, was primarily interested in a moral order of the universe. But it is fascinating to observe how closely he followed the received astronomy and physics of his time. The center of Earth was the center of the Aristotelian universe, and it is the center (and the bottom) of Dante's as well: that point to which all heavy (sinful, earthy) matter is attracted. It is there that the creator of sin, Satan, is forever transfixed (see Figure 26.2).

Shortly, Dante and Virgil will begin their difficult climb down Satan's body (clinging to his hairy shanks) to descend to the very bottom of Hell (the center of the universe) and, having passed this point, begin the equally difficult ascent to Purgatory. (Please excuse the translation.)

From shag to shag he now went slowly down,
Between the matted hair and crusts of ice.

[1] Alighieri, D. (ca. 1321). *The Divine Comedy*, transl. Lawrence Grant White, New York: Pantheon Books, 1948, Canto 34.

Fig. 26.1 *The Ninth Circle of Hell* by Gustave Doré.

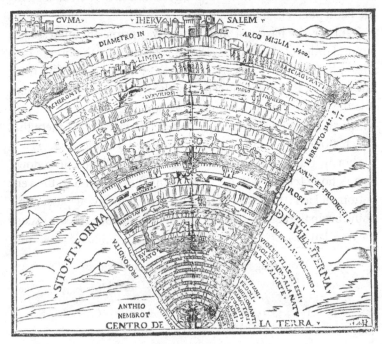

Fig. 26.2 Dante's Geography of Hell.

> When we had reached that point just where the thigh
> Does turn upon the thickness of the haunch,
> My leader, with fatigue and labored breath
> Brought 'round his head to where his legs had been,
> And grasped the hair like one who clambers up,
> So that I thought our way lay back to hell.[2]

Virgil explains:

> ...when I turned around, we passed the point
> To which all weights are drawn from everywhere.[3]

In a Newtonian universe, their experience would have been more like that of an astronaut: they would have floated. (At the center of Earth, the gravitational force goes to zero.) This only reported journey to Earth's center seems to favor Aristotle. But, regrettably, all of our other experience favors Newton and his great descendant Einstein.

The second story begins possibly before the first. But let us start in Greece of Pericles. It was known then that amber (electron) when rubbed attracted chaff. As Plutarch was to put it:

In amber there is a flammeous and spirituous nature, and this by rubbing on the surface is emitted by hidden passages ... [4]

After two millennia of scientific progress Gilbert noted:

Alexander Aphrodiseus incorrectly declares the question of amber to be unsolvable, because that amber does attract chaff, yet not the leaves of basil ...

And he adds testily:

...but such stories are false, disgracefully inaccurate.[5]

2 Ibid.
3 Ibid.
4 Plutarch, *Morals*, translated from the Greek by several hands, corrected and revised by William W. Goodwin, with an Introduction by Ralph Waldo Emerson, 5 vols. Boston: Little, Brown, 1878.
5 Gilbert, W. (1600). *De magnete*, transl. P. Fleury Mottelay as *On the Lodestone and Magnetic Bodies*, New York: John Wiley, 1893.

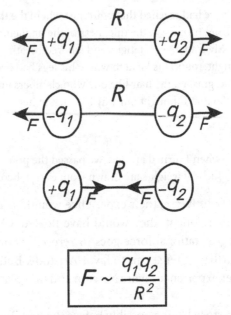

Fig. 26.3 Coulomb's law of attraction and repulsion of electrical charges.

Almost two thousand years of such deep thinking came together in the seventeenth and eighteenth centuries with contributions from Luigi Galvani and Alessandro Volta resulting in several key ideas: the source of these attractions is electrical charges which come in two varieties: positive and negative. These repel or attract according to Coulomb's famous law (see Figure 26.3).

Soon thereafter Oersted and Ampère formulated the laws of magnetic forces – a special glorious complication of electrical systems – these forces depend on the relative motion of the charges (and go to zero if the charges are at rest with respect to one another) (see Figure 26.4).

With these discoveries, in the early nineteenth century, a frenzied search began to produce electrical forces from magnetism. Success was finally obtained, almost simultaneously, by Michael

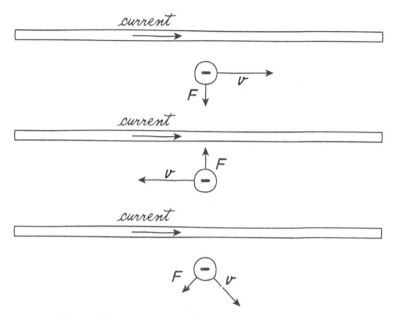

Fig. 26.4 Magnetic forces due to currents or charges in motion.

Faraday in England, M. F. E. Lenz in Russia, and Joseph Henry in the United States. It was changing magnetic fields that produced electric forces, a discovery that led to our present means of generating electricity (see Figure 26.5).

Margaret Thatcher once commented in a talk to the Royal Society, that the value of Faraday's discovery was greater than the total capitalization of the London Stock Exchange (having, no doubt, reduced the budget for fundamental research the day before). And Faraday himself when asked by then Chancellor of the Exchequer William Gladstone about the practical worth of electricity answered, "One day, Sir, you may tax it."[6]

---

6 Faraday, M. (1918). Quoted in Mackay, A. L. (1977). *The Harvest of a Quiet Eye: A Selection of Scientific Quotations*, Bristol: The Institute of Physics, p. 56.

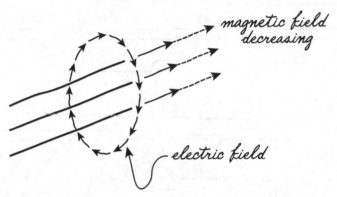

Fig. 26.5 Changing magnetic fields produce electric fields.

James Clerk Maxwell now enters the scene – Maxwell, who is presently placed with Newton and Einstein at the summit of the Pantheon of theoretical physics, writes:

I have endeavored to make it plain that I am not attempting to establish any physical theory of a science in which I have not made a single experiment worthy of the name ... [7]

Modesty I have not noted in recent statements of my theoretical colleagues.

Maxwell attempted to codify all that was known of electrical and magnetic phenomena in a concrete, consistent, and economical set of rules: a few lines of mathematical abstractions.

As Faraday was to write:

I was at first almost frightened ... when I saw such mathematical force made to bear upon the subject, and then wondered to see that the subject stood it so well. [8]

In the course of this effort Maxwell discovered that the rules culled from the experiments of Coulomb, Oersted, Ampère, and Faraday were inconsistent. We could spend an entire lecture discussing how

7 Maxwell, J. C. (1865). Letter to Lord Kelvin.
8 Faraday, M. in a letter to Maxwell.

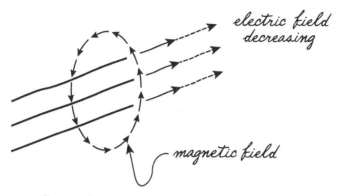

electric field decreasing

magnetic field

Fig. 26.6 Changing electric fields produce magnetic fields.

this could come about, but let us just say briefly: incorrect extrapolation or, in Maxwell's words:

> We must recollect   that no experiments have been made … except (for) closed currents … [9]

To produce consistency Maxwell modified one of the rules (the relation between moving charges and magnetic fields) by adding an additional term that resulted in an additional possibility: changing electric fields produced magnetic fields (see Figure 26.6).

Now a miracle: the modified rules reproduced what had previously been observed by Faraday and others. But they contained some stunning and unprecedented new possibilities. Previously electric and magnetic forces could exist only in the presence of fixed or moving charges. Now Maxwell found that such forces (fields) could sometimes exist *in the absence of charges.*

They come in the form of waves of all possible wavelengths, waves of alternating electric and magnetic fields propagating through space at a speed he could determine using measurements made in previous electrical experiments.

9 Maxwell, J. C. (1861). On Physical Lines of Force, *Philosophical Magazine* 4th series, four parts: I, vol. 11, pp. 161–175; II, vol. 11, pp. 281–291, 338–347; III, vol. 13, pp. 12–23; IV: vol. 13, pp. 85–95.

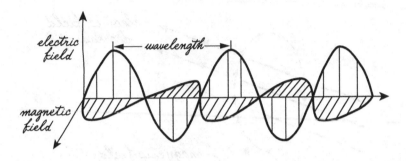

The following are handwritten annotations within the figure:

electric field

wavelength

magnetic field

• all wavelengths are possible

• in a vacuum all wavelengths travel at the same speed -- the speed of light, about 3×10^{10} cm/sec

Fig. 26.7 Electromagnetic waves.

"The velocity of transverse undulations in our hypothetical medium," he writes, "calculated from the electromagnetic experiments of MM. Kohlrausch and Weber, agrees so exactly with the velocity of light calculated from the optical experiments of M. Fizeau, that we can scarcely avoid the inference that *light consists in the transverse undulations of the same medium which is the cause of electric and magnetic phenomena.*[10] (See Figure 26.7.)

And, in a letter to William Thomson (later Lord Kelvin):

I made out the equations in the country before I had any suspicion of the nearness between the two values of the velocity of propagation of magnetic effects and that of light, so that I think I have reason to believe that the magnetic and luminiferous media are identical...

Further:

...it seems we have strong reason to conclude that light itself (including radiant heat, and other radiations if any) is an electromagnetic

10 Ibid.

disturbance in the form of waves propagated through the electromagnetic field according to electromagnetic laws.[11]

The importance and practical consequences of the identification of light as an electromagnetic wave cannot be exaggerated. All of the properties of light: reflection, refraction, transmission, as has been worked out in detail by several generations of physicists including Hertz, Fitzgerald, and Lorentz now follow as a consequence of known electrical interactions, interactions of electric and magnetic fields with charged particles. And this led, very soon, to Albert Einstein's profound redefinition of time itself.

But conceptually what Maxwell did is perhaps more astonishing and profoundly illuminating as to the nature of science.

What does electricity and magnetism have to do with the ineffable essence of that independent category or entity "light," by explicit command of the Almighty, third in the order of creation. In what sense can light be an electromagnetic wave? The answer is science at its most beautiful and unapologetically reductionist.

If we replace the word "light" in all sentences by the possibly less sonorous and more cumbersome "electromagnetic wave" we have the equivalent in logic (not in sound, poetry, or psychological evocation). There is no need for the concept of light as a separate primitive. Thus the ineffable has been reduced: constructed from more primitive elements. We note that neither the value nor the significance of light have been diminished.

If time allowed, I would love to regale you with some of the very many astonishing and beautiful relationships that follow. But let me bring us to the present following a single magnificent path.

Maxwell published the first extensive account of his theory in 1867. Eight years after his death, in 1887, Heinrich Hertz produced and detected Maxwell's electromagnetic waves.

11 Maxwell, J. C. (1865). Letter to Lord Kelvin.

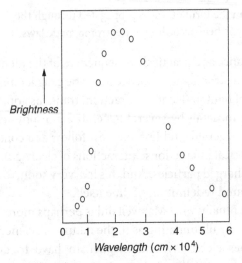

Fig. 26.8 Observed black-body radiation at a temperature of 1600 kelvins
(1327° C).

A decade later, another physicist, Max Planck, pondered long and hard on the properties of what is known technically as black body radiation, but for our purposes can be thought of as the light coming from the grill on the top of an electric stove. It is common experience these days that as the grill warms up its color changes from dull red to brighter red and if hot enough might even look whitish (white hot).

Stated somewhat technically, Planck was trying to understand the color distribution of Maxwell's electromagnetic waves when in the thermal equilibrium (black body radiation). Experimentally, the situation looked as in Figure 26.8.

Notice first that there is a very specific relationship between brightness and color. Notice also (although it is not necessarily evident in Figure 26.8) that the maximum intensity of radiation occurs at different colors at different temperatures (going toward violet as temperature increases). Thus the grill (or the black body) looks bluer at higher temperatures. Further, this relationship holds

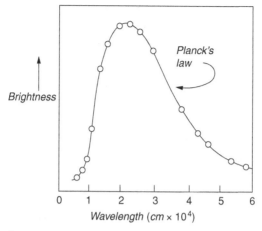

Fig. 26.9 Comparison of Planck's law and the observed black-body radiation at a temperature of 1600 kelvins (1327° C). The white circles are the observations.

even if *no visible light comes out of the object.* (The visible region is but a small portion of the electromagnetic spectrum.)

It is incidental to the story I am telling (though it does lead to a monumental event in the history of science) that existing physics seemed totally incapable of explaining these curves. The predictions of then current physics were wildly different from what was seen – and in fact, seemed totally paradoxical – leading to what became known as the ultraviolet catastrophe. (In those days, long before gluons, physicists used words as well as mathematics with a certain *élan.*)

To resolve this dilemma, Planck introduced the idea of the quantum, initiating one of the major intellectual revolutions that ushered in the twentieth century; but more important for him at the time, obtaining theoretical curves that reproduced the experimental data very well (see Figure 26.9).

It is now 1963. Arno Penzias and Robert Wilson have obtained use of a large horn-shaped antenna (designed for communication with satellites) for a foray into radio astronomy, searching the

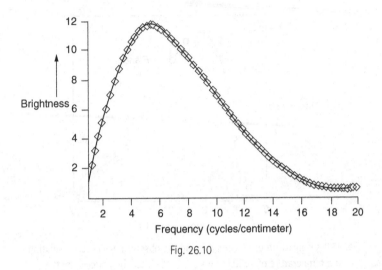

Fig. 26.10

heavens for electromagnetic wave emissions in the microwave wave region.

When listening they heard annoying static in the background. And very much as you or I might try to tune in to a favorite station, they went to great lengths to eliminate this static. After many heroic attempts, they finally come to the realization that the static was the music.

This electromagnetic radiation they were detecting (in the microwave range) came with extraordinary uniformity from all portions of the heavens and when they combined their results with those of others realized that its intensity varied with wavelength just as would be the case for a black body.

Figure 26.10 is obtained from the recent remarkable observations of COBE, the Cosmic Background Explorer Satellite.

The squares are the observed points while the solid curve is the theoretical distribution of radiation from a black body at a temperature of 2.7 K. That is 2.7° above the absolute zero.

Where does this come from?

It turns out that such black body background radiation had been predicted by Gamow in 1948 as consequence of an expanding universe that resulted from a big bang.

The story as we tell it today goes as follows.

About 100 000 years after the creation when the temperature of the universe was about 3000 kelvins, the universe became transparent to electromagnetic radiation (transparency occurred when the temperature was low enough so that electrons would be captured by protons to form atoms; I don't have the time to explain why).

From this time on, the entire universe could be regarded as a black body, the electromagnetic radiation in equilibrium (thus following an intensity wavelength curve just as the ones I have shown). As the universe expands the temperature of the enclosed radiation drops according to a known relation till today (about 13 billion years after the creation) the temperature should be of the order of a few degrees above the absolute zero (the exact temperature measured by the curve on the slide is one determination of the age of the universe).

I hope this sketch suggests the span of the structure we have created: from amber to light to the origin and age of the universe. Thousands of detailed observations are in agreement, often to startling numerical precision, with what is expected. In the area of the quantum theory of Maxwell's electromagnetic waves, known as quantum electrodynamics, there are no known disagreements.

But further, the story illustrates how that ill-reputed process of reductionism actually works. When light is "reduced" to electromagnetic radiation, it is in no way diminished. But it is no longer necessary as an independent logical primitive, very much as in geometry a triangle is constructed from lines, no separate logical primitive "triangle" is required.

Now the third story – a story not yet complete – an ongoing effort that for more than twenty years has involved psychologists, neurophysiologists, mathematicians, and physicists.

The object is possibly grandiose – to understand the brain – from the physical or molecular basis for the most fundamental operations such as learning and memory storage to system properties: how information is stored, processed and/or recalled and how at the highest level this system finally becomes aware of itself, conscious. Certainly an ambitious project, in some aspects more advanced than is generally realized.

But before I discuss some of the progress in the area of fundamental mechanisms, let me attempt to address a concern that may immediately trouble some of you.

How could one, from ordinary materials, construct something as ineffable as our mind, our consciousness or, if we wish, our soul (mortal or immortal). How would we know that the constructed system was conscious? Is this even a scientific problem? Although I will attempt to address such deep mysteries later, for the moment beware of the modern Alexander Aphrodiseus who tells us that, "the problem is unsolvable." Possibly because we are more conscious of chaff than basil leaves.

In the early 1970s much was known about the properties of neurons, nerve cells, but almost nothing about how systems of such cells could store and process information. The neuron is a very complex cell that is almost universally believed to play an extremely important (if not the entire) role in the transfer and storage of information in the brain. Neurons transmit signals electrically (generally FM signals). These are passed from cell to cell chemically by precise mechanisms that are very complex but fairly well understood. But for our purposes think of a neuron as in Figure 26.11.

Memory is now thought to be stored in modification of synaptic junctions and various theories have been developed to describe how this comes about. I don't have the time to go into details,[12] but I believe that it is reasonable to say not only that we have made more progress than is generally appreciated, but, given the level

12 Some of these are given in Chapter 17, "Memories and Memory: A Physicist's Approach to the Brain."

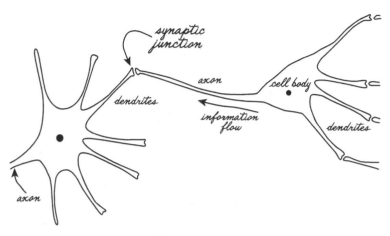

Fig. 26.11 Idealized neurons.

of skepticism displayed when ideas such as synaptic modification were talked about twenty years ago, that we have astonished even ourselves. The field has changed. It is no longer uncommon to have an experimental paper with a possible neural network explanation. Nor is it uncommon to hear elaborate biochemical mechanisms proposed as the basis for memory storage.

Permit me then to project into the future with a certain optimism. Assume that the fundamental mechanisms are understood, that the neural correlates of mental states (as Francis Crick is seeking) are in hand. Do we then understand the human mind and its presumed origin in that biological organ, the brain?

Although much emphasis has been placed on algorithmic behavior or what are very loosely referred to as neural computational processes, these properties in my opinion – I don't have the time to expand here – will probably be easier to understand than such properties as feeling, awareness, or consciousness. Although it has been suggested by some that these latter properties will arise as a consequence of the execution of the proper algorithmic behavior, my opinion is that algorithms and consciousness are relatively independent. After all, hand-held calculators execute algorithms and little dogs wagging their tails don't do much arithmetic. Perhaps we can phrase the question as follows: What are the steps required so that a machine could experience mental activity? Could we, in the extreme, construct a machine that was conscious?

What we must understand is how consciousness arises as a property of a very complex physical system. It is this, not reasoning, that in my opinion remains the profoundest mystery surrounding that biological entity: the brain.

The scientific problem, as I see it, is to construct from material components such as neurons and/or systems of neurons the simplest entity that performs the most primitive conscious act. (To paraphrase Bourbaki, a beautiful problem, it can be stated very simply and will, no doubt, have a very complex solution.)

A satisfactory theory of mind not only is allowed, but, in my opinion, requires the introduction of mental entities. We will be satisfied only when we see before us constructs that can have mental experience, when we see how they work, how they come about from more primitive entities such as neurons.

It is possible, in my opinion, that there is no sharp demarcation between the categories conscious and non-conscious (just as we would say today that no such sharp distinction exists between the categories living and non-living). It could even turn out (as shocking as this might appear to my scientific colleagues) that we *must* invoke a new "law of nature," follow Descartes and pour the conscious substance into the machine.

But the conservative scientific position is to attempt to construct this seemingly new and surely very subtle property from the materials available, those given to us by physicists, chemists, and biologists (as has been done many times before: celestial from earthy material; organic from inorganic substances; the concept of temperature from the motions of molecules; or, as in the second story, light from electricity and magnetism). The unrepentant reductionist believes that this construction can and will be made; that it will in no way diminish the value or significance of what has been constructed; that, to paraphrase Santayana,

All our sorrow is real, but the atoms of which we are made are indifferent.[13]

Success would no doubt be magnificent but failure might be more so. If we cannot perform the reduction then we will genuinely have made one of the most profound discoveries in the history of thought, consequences of which would shape our conception of ourselves in the deepest way. And, in spite of the winds of current political fashion, might even be relevant.

I have attempted with these three stories, the first and second already encased in Literary and Scientific Halls of Fame, the third sure to be elected if and when we succeed, to celebrate the very difficult work we do in research and scholarship and, as important, the concomitant communication of what we know and what we do to others that will follow: to celebrate what has been done here and elsewhere for nine hundred years.

But we are under attack. There are those who question not only the worth, but also the validity of what we do. Can science reach so far? Are there any limits to what intellect can understand? In every era there is criticism as well as doubt. Presently we hear talk of chaos and complexity just as in the recent past we heard of catastrophe theory; to my mind these are old arguments in not-so-fashionable new clothing. We are told, for example, that there may be an overall theory of all complex systems. I suppose there may be, but for those of us in the trenches most different complex systems seem very individual. In spite of the fantasies of certain elder statesmen, there appears to be no substitute for the hard work of getting to know each system in order to analyze it.

As for chaos and computability, we are not particularly surprised that some events either because of the number of variables involved

---

13 Santayana, G. (1925). *Dialogues in Limbo*, New York: Charles Scribner's Sons.

or because of their sensitivity to initial conditions are, as a practical matter, not possible to compute. We are not shocked to learn that it will be difficult to compute whether or not a well-balanced penny will land on heads after it has been shaken for a few minutes in a box. What is really surprising is that so much that is very important is not chaotic and is computable.

And for the lately fashionable new academic left that characterizes science as dominated by white male ideas, they have merely to present to us an equally coherent structure in equally remarkable agreement with what we observe and we will begin to take them seriously.

You will gather that these criticisms do not trouble me very much. My personal recommendation is not to pay too much attention, not to get too excited. Just as the 1920s American college fashion of swallowing goldfish (of which you may have heard), they will go away.

But it is legitimate to ask simply: In spite of our remarkable even astonishing success, are there limits? Are some things intrinsically unknowable, say the state of the universe before the Big Bang? To this question, we don't know the answer. But before we leap to conclusions, let me remind you that early in the nineteenth century Auguste Comte gave as an example of the intrinsically unknowable the chemical composition of the stars.

To this deepest of questions not much has been added, in my opinion, since it was so beautifully put over two thousand years ago in the contrasting views of Classical Greece and the Old Testament. The Greek tradition dating from Thales:

Nothing in nature is permanently hidden from the human mind.

The Book of Job, expressing the Judeo-Christian tradition of inaccessible mysteries, God speaking to Job out of the whirlwind (if I may be bold enough to paraphrase):

How can you be so pretentious to believe that with your pea size brain you could ever understand what I have done?

In fact, those of us who are scientific optimists must regard it as somewhat of a happy accident that our brain, arising out of nature, is large enough to construct scientific structures sufficiently interesting to motivate us to continue. And since we have progressed to the point at which we can construct auxiliary thinking machines, there may be no limit. It is my personal hope that the Greek faith is correct, that in nature there are no dark corners permanently inaccessible to the light of human intellect.

To my mind the threat that science or rational human intellect faces lies not in sometimes fashionable, foolish criticism or loose talk about limits, but perhaps in the emotion that lies behind this criticism (those hard-to-control emotions, surely as much part of ourselves as our intellect, our non-rational self; what makes life delicious as well as hideous) allied with our enslavement to outworn ideology. For science, although it does not dictate what we may believe, often makes certain of our most cherished beliefs unfashionable. And people would rather be wrong than unfashionable.

When we consider that we cannot agree on divisions of territory; that ethnic strife goes on for ever and gets worse; that we cannot agree on so simple a proposition as eventually what each of us can enjoy, in a world driven to equality of living standards, is the ratio of total resources divided by total population; that we cannot go back to nature, a nature that never was, but in any case a nature no longer accessible since there are so many of us; that on a planet with limited resources what we need desperately is to hold our numbers more or less constant while we dig ourselves out of the mess we have made.

Instead (isn't it borderline scandalous) old men with vast authority, long insulated from life's vicissitudes, who ages ago have forgotten the anxiety, passion, loneliness, and unfulfilled desire of youth, issue proclamations on how youth should behave, that youth must deny what biology has driven them to desire most to solve the most pressing problem this planet faces.

Let us then celebrate nine hundred glorious years of continuous intellectual activity; research, scholarship, and, as important, the communication of our tradition to those that will follow.

For the future there is no greater promise than this magnificent heritage, no greater burden than the heavy weight of outworn ideology. The threat lies not in possible limits of science or dark corners of nature, but rather in possible limits or actual dark corners of our own soul.

Printed in the United States
by Baker & Taylor Publisher Services